D0031480

The Digital Rights Movement

The Information Society Series

Laura DeNardis and Michael Zimmer, Series Editors

Interfaces on Trial 2.0, Jonathan Band and Masanobu Katoh

Opening Standards: The Global Politics of Interoperability, Laura DeNardis, editor

The Reputation Society: How Online Opinions Are Reshaping the Offline World, Hassan Masum and Mark Tovey, editors

The Digital Rights Movement: The Role of Technology in Subverting Digital Copyright, Hector Postigo

The Digital Rights Movement

The Role of Technology in Subverting Digital Copyright

Hector Postigo

The MIT Press
Cambridge, Massachusetts
London, England

MIT Press books may be purchased at special quantity discounts for business or sales promotional use. For information, please email special_sales@mitpress.mit.edu or write to Special Sales Department, The MIT Press, 55 Hayward Street, Cambridge, MA 02142.

This book was set in Stone Sans and Stone Serif by Toppan Best-set Premedia Limited. Printed and bound in the United States of America.

Library of Congress Cataloging-in-Publication Data

Postigo, Hector
The digital rights movement: the role of technology in subverting digital copyright / Hector Postigo.
p. cm. — (The information society series)
Includes bibliographical references and index.
ISBN 978-0-262-01795-4 (hardcover: alk. paper)
1. Copyright and electronic data processing. 2. Digital rights management. 3. Hacktivism. 4. Internet—Law and legislation. 5. Piracy (Copyright)—Prevention. 6. Fair use (Copyright). I. Title.
K1447.95.P67 2012
345'.02662—dc23
2012004559

10 9 8 7 6 5 4 3 2 1

Contents

Part I

1 Introduction

In October 1999, a small group of hackers[1] developed the program DeCSS (for "Decrypt Content Scrambling System") to crack the encryption system on commercial DVDs and posted the software and its code on the Internet, distributing it worldwide. The DeCSS source code and the DeCSS application served as tools for those individuals designing DVD players for computers running on the Linux operating system. Because all DVD players must have a way of decrypting the information on a DVD before they can play the movie, DeCSS was invaluable in developing early DVD player technology for computers using operating systems other than Windows or Mac OS (Warren 2005).

The DVD Copy Control Association (a consortium of copyright interests such as movie studios who license CSS), following the release of DeCSS in 2000, mounted a legal campaign against Internet sites publishing the DeCSS code, distributing the application, or linking to sites distributing the application and code. They argued that DeCSS violated the Digital Millennium Copyright Act (DMCA) of 1998[2] by allowing the circumvention of technology designed for copyright protection and by promoting unsanctioned copying and distribution of protected material.

Despite mounting legal pressure, supporters of DeCSS started a legal campaign of their own, arguing that as owners of the content on DVDs they should have access to those data and be allowed to make copies for personal use. Furthermore, some DeCSS supporters mounted a campaign of civil disobedience in defiance of court orders to remove the DeCSS code from their Web sites. One such activist, David Touretzky, argued that the court sanction was a violation of his right to free speech and posted a gallery of CSS descramblers. On his Internet site, he made available the CSS descrambling code in verse form and as a recording of a person singing the descrambled code to music (Touretzky n.d).

Examples of hacks against copy-protection/access-protection technologies and mobilization against a host of regulations and business practices that limit consumer access and use over legally purchased cultural products have become common since the days of the DeCSS controversy. These types of activism challenge long-held industry and legal perspectives on what the roles of users and media consumers are in relation to the products produced by the cultural industries. This book undertakes a historical analysis of legislation and case studies that demonstrate the origins, themes, and structure of digital rights activism as it emerged in the late 1990s and early 2000s. The analysis points to a coordinated movement that seeks to ensure a culture of participation in media products: what I call the "digital rights movement."

So far as social movements go, the digital rights movement is not especially well known among broader publics—not in the same way as, for example, antiglobalization movements that have made headlines in recent years. It is, however, a movement nonetheless and one that is of increasing importance to a broad base of new and old media consumers. In short, the movement is a concerted effort to ensure the rights of consumers and users of digital media and technology. The issues generally addressed include privacy, free speech, fair use, technological innovation, and first sale.

The struggle between digital rights activists and the content industry is novel for a number of reasons. First, it is highly technological, meaning that it is dependent on technology at least in part to implement some of its collective-action goals and to realize the kinds of social change it seeks. Furthermore, for the movement, digital technologies such as computer programs, the Internet, and media hardware are *both* the obstacles it faces as well as the means it uses in resisting/undoing the constraints on consumer use and access. Second, we should note the contingent nature of the term *digital rights* and point out that it refers to a broad set of practices that are not always or necessarily "digital." Therefore, if we speak more broadly, the digital rights movement is concerned with culture (mass-produced culture) and control over its production.

In its analysis of the digital rights movement, this book addresses a number of tasks. First, it revisits the early legislative history of the DMCA, illustrating the policymaking process and showing its discursive construction and how lawmakers and content industry representatives in the 1990s imagined the World Wide Web (what was then called the National Information Infrastructure) and consumers therein. These imaginaries represent visions of the kind place the Web would become, the kinds of consumers who would traverse it, and the kinds of technology needed to make it run smoothly. Blind spots in these imaginaries ultimately yielded laws (the DMCA in this case) that would be at odds with existing and emerging consumer practices. The historical analysis shows that the DMCA's formational discourse is the discourse of US copyright law, so the imaginaries deployed in moments of deliberation during the DMCA's formulation also reflect the rationalizing rhetoric of copyright writ large and its visions of cultural production. The DMCA, then, can be read as an instrument of the copyright statute in the United States (and abroad), bringing its rationale to bear on digital media, the Internet, and other digital technologies. This act and by extension copyright law are the laws in whose name many early prosecutions and lawsuits spurred activism in the case studies discussed in subsequent chapters.

Part II deals with case studies that are related to the DMCA and issues that activism against increased control over digital media have brought to light. The case of DeCSS is chronicled in some detail as are eBook hacks, iTunes hacks, and other forms of hacking orchestrated not only by single hackers, but by activist organizations. This second part is related to the first in that it shows how resistance to the outcomes of the legislative process took shape early on and how that resistance brought to light important issues for the movement, such as user-centered notions of fair use, free speech, and a discourse of consumer rights over content that are often bargained away in click-through agreements.

In the remainder of this introduction, I discuss the implications, issues, and themes related to the events and topics discussed throughout the text.

What Does It Mean to Think about a Digital Rights Movement?

This work is necessarily historical (though not exactly a history) and should be read as a picture of the digital rights movement as it was when it first began to coalesce and take action (primarily against the DMCA and its excesses). When I first started writing about it in 2006, it seemed very much a movement about consumer rights in digital content, concerned primarily with the technological impediments to digital media consumption and the laws that abetted them. But today the movement is more than that. Activists have started referring to themselves as part of a free-culture movement, for example, because what started off as an awareness of the limitations imposed on consumer access and use of mass-cultural products parsed through digital media has become an awareness of increasingly stringent laws and technological measures that lock up access to the "cultural commons." It seems to me that the movement no longer pivots on what technology and associated policy can or cannot allow in terms of consumption but now focuses on culture and what people believe access to cultural production (not just consumption) should entail.[3] The movement today is as much about cultural change (a change to a culture that is participatory) as it is about legal and technological change and digital rights.

Not long ago legal scholar James Boyle presciently wrote about the possibility of such a movement. Commenting on the changes in intellectual property brought on by technological change, Boyle suggested that perhaps we are in need of a politics of intellectual property to protect the public domain from what he termed a "copyright land grab" (1997). He described this "land grab" as driven by new technological affordances present in emerging systems for distributing copyrighted works and for control over their use. When he wrote about these issues (fair use, participation, access to content, and the cultural commons in the digital world) in 1997, the fight over digital rights was just emerging, and he noted that those issues seemed to be fractured and affecting divergent populations ("software engineers, libraries, appropriation artists, parodists, etc."). Boyle suggested that what the various stakeholders needed were "analytical frameworks" that would bring them together and address what appeared to be the inexorable logic of the current system.[4] The analytical frameworks he suggested included first a critique of the failure in decision-making processes in formulating copyright law that gives the pretense of benefitting society but really ultimately benefits few and passes on the costs of failures in the system to the whole of society, an appropriate critique given the legal debates over copyright at the time he wrote his essay.[5] The DMCA had just been formulated, and many legal scholars were starting to see, because of emerging case law and the policy process, that increased legal protections tied to technology measures were seriously endangering the public domain.

Boyle also proposed a critique of our (Western, US-based) concept of intellectual property as foundational for organizing intellectual products. He was especially critical of what he termed the "original author" concept, which he argued turns a blind side to the cultural commons from which such authors must draw. Most important, perhaps, he pointed to the need for a convincing rhetoric of the politics of intellectual property and the cultural commons, one that draws in not only directly vested actors, but also those who may not necessarily have considered themselves to be affected by the issues.

It seems that this process of formulating viable critiques is well under way today as activists have coalesced into a recognizable movement. For example, in 2009 the Free Culture Forum organized by Exgae, Networked Politics, and the Free Knowledge Institute, three organizations working on digital rights issues, was held in Barcelona and gathered together a diverse host of activists, academics, and others from across western Europe and the United States. I was there as an observer, and I was interested in the conference for what it had to say substantively about the state of digital rights and how the discourse had changed. The migration of most of the cultural industry's products to digital media, the rise of a participatory ethos among the young, and the ever-increasing technological affordance and impediments to access cultural goods made the conversation more global. Now activists were demanding rights to access and use cultural products as well as to participate in production. These demands had even greater import when framed by long-standing debates over commercialization, mass production, and privatization of mass culture in its various forms.

What began in the United States as a debate over the acceptable limits of copyright in the digital age has morphed into a global debate about the acceptable limits of law in safeguarding cultural products for large corporations. Debates about net neutrality, copyright, digital rights management, and participatory audience practices are in essence debates about cultural ownership. Increasingly throughout the modern/modernizing world organizations, intellectuals and all manner of activists are weighing in, trying to articulate a number of "participatory rights" never before expressed by consumers.[6] So one of the first things we can say when thinking about the movement is that its core goals make it more expansive in its impact than we might at first see. The movement is not about consumers consuming and the gadgets they need, but rather about developing a legitimating discourse in law and technology for participation in cultural production.

My conclusions about the meanings and means of the digital rights movement are: (1) activists, intellectuals, and organizations in the movement call for a culture that is participatory in mass-cultural products (requiring the ideological, legal, and technical affordance to realize such a culture); and (2) the means for achieving this culture are, as one would expect, institutional and extrainstitutional. This means that activists seek legal change both through traditional political venues such as the legislature and the courts as well as through nontraditional means such as protest and other forms of direct action.

On this last point, it bears pointing out that part of the extrainstitutional repertoire includes the design and distribution of technologies meant to counteract the effects of existing technolegal regimes (laws and technologies that regulate user practices). These activities amount to more than hacking or "hacktivism" as it has been traditionally understood (Jordan 2002; Jordan and Taylor 2004) because such design practices are explicitly political, the technologies are explicitly meaningful (not just instrumental), and their presence in the ecology of resources available to the movement empowers the movement and individuals within it beyond what has traditionally been possible.

To put the second point more concretely, the practice of designing and distributing technologies that may, for example, circumvent copy-protection measures or work around existing paradigms for content distribution can be carried out by *individuals* and is not limited to organizations (a point that in itself is significant). So where once these kinds of impactful tactics would require large organizational resources, the possibility that a lone hacker can release a powerfully disruptive technology that is potentially widely adopted decenters the social movement organization (SMO) as a keystone for powerful collective action. More important, however, *the material presence of such technologies realizes the world they seek.* In other words, technologies such as those briefly introduced earlier in connection with DeCSS and discussed in later chapters of this book serve a double function. Their creation and existence can be read as a form of protest (so they are meaningful beyond their function), but they also realize part of the central goal the movement seeks: a culture that is participatory (with the tools to engage in participation).

Imagine as an analogy a movement like the one that led to the Americans with Disabilities Act[7] in the United States, which required mobilization by persons living with disabilities. One outcome was that our cities' street curbs were redesigned to accommodate wheelchair access. This outcome required resources and extensive petitioning to city, state, and federal agencies. But imagine if those activists had circumvented the state and its resources and simply gone out and altered the curbs themselves both as an act of protest and as a way of realizing the world they sought.[8]

Designing technologies in the digital rights movement has the latter powerful effect. It allows for the creation of a parallel technological architecture and eventually parallel technolegal architecture when paired with changes in law or new licensing practices such as Creative Commons. For those familiar with Lawrence Lessig's work to develop the Creative Commons licensing scheme, one cannot help but see it as an elegant hack. Whereas some were busy hacking the technologies that prevented a culture that is participatory, Lessig and others hacked the licensing practices that worked in tandem. Thus, another important point that becomes evident when thinking about the movement is that hacks matter in a structural and meaningful way: they have impact on the structure of activism; they have impact on the structure of consumption; and they have impact on the normative power of law.[9]

It should strike some readers how much this idea of participating in culture is like Henry Jenkins's concept of participatory culture. However, what I propose herein is not necessarily about participatory culture. I suggest that the movement's understanding of culture as participatory is subtly different from Jenkins's concept of participatory culture in studies of fandom and more recently in his and others' accounts of convergence culture, where an increasing number of consumers behave like producer and consumer at the same time (Banks 2005; Hartley 2006; Kucklich 2005; Postigo 2007, 2008). Specifically, the concept of participatory culture from Jenkins and others speaks of a culture of participation among subsets of content consumers. I would contrast this view of a participatory culture to the digital rights movement's notion of culture (the whole of shared meanings parsed through mass media and new digital technologies) as *necessarily participatory*. Culture for the movement is meaningless or increasingly alienated from a citizenry *unless that citizenry can participate in its production*. To understand the relationship between Jenkins's participatory culture and the movement's definition of culture as participatory, one might think of participatory culture as one of the means by which culture writ large may become participatory (other means might be legal or technological, formalized into the workings of society by institutions). It may be the case that the practices of participatory culture may someday be widespread enough that they become the way consumers see their relationship to mass media and the mass-media experience—they will see their hand in the products of the cultural industries. In that case, the two concepts—participatory culture and culture that is participatory—might converge. For now, they remain related but different.

Themes Explored in the Book

The Meaning of Fair Use and Related Legal Concepts

A key concept for the movement is "fair use," a legal concept first and foremost, but importantly for the movement in the United States a discursively powerful springboard for arguments about rights (participatory, creative, digital, cultural).

Twenty years ago the term *fair use* was not part of the popular vernacular. Teenagers and college students did not know and discuss fair use;

concerns over the particulars of fair use were the worries only of university information officers and librarians. Today, fair use has a pressing need to be understood by a broad number of publics. In 2005, the Supreme Court heard for the fifth time in its history a case where fair use was a defense for potential infringement of copyright (MGM v. Grokster et al. [545 US 913 (2005)]). The possibility of easily copying, distributing, publishing, and performing copyrighted content in digital formats has made fair use a real concern for both copyright owners and consumers of copyrighted material.

Although statute and precedent have established an approach for judging the merits of claims of fair use, it is an important concept beyond the strict confines of its statutory definition. As the digital rights movement took shape, activists conceived fair use in a user-centered fashion. Their interpretation of fair use sought to legitimize personal noncommercial uses (such as making back copies of songs) and noncommercial creative uses (such as remixing music and video tracks). When activists challenged the DMCA's anticircumvention provisions in court, they considered fair use to be a tool for ensuring free speech.[10] SMOs, hackers, and other activists fought to capture free speech and fair use as representative values for the movement. Framing the digital rights movement as a movement for free speech and fair use was a key strategy because it positioned the movement's goals within accepted and cherished values in US society.

Copyright owners also deployed their own framing strategy, however, portraying hackers as criminals; fair use as a privilege, not a right; and the balance of copyright as sacrosanct. In many ways, fair use has been popularized by the prominence of this debate. The struggle between competing frames is the background over which the technologies that protect copyright and those technologies that circumvent protection clash in what has aptly been called a "code war" (Biegel 2001).

Fair use, then, is what social movement theory would call an important "master frame" in the digital rights movement. It conceptually brings together ideas that emerge about access and use of cultural products, creative rights, and participation into a narrative that can be ported beyond the movement to other publics. In many ways, the idea of fair use has allowed the movement to grow beyond its initial confines of digital rights to arguments about free culture.

Technology as Enforcement

Lawrence Lessig and others have pointed out the role that code plays in regulating or acting as a surrogate/partner for enforcing legal regimes. In this book, the concept of technology or code as law is further explored, but with an eye toward its meaningful place as an obstacle for the movement. In other words, the insights of viewing code and any technology as potentially regulatory are not rehashed here for what they say about society as a whole but rather for what they say about what the movement must confront. Put more succinctly, the digital rights movement, unlike many other social movements, confronts not only legal regimes, but technological regimes as well, some of which exist outside the reach of traditional institutional mechanisms for social change (lobbying, for example).

In Lessig's oft-cited model of the regulatory power of code (Lessig 1999), the individual is seen as a dot at the center of four modalities: norms, the market, law, and architecture or technology. With each modality exerting pressure on the individual, behavior is a result of the sum of the various pressures. Lessig noted that the dot was "pathetic" because its actions were at the whim of these modalities. Technological enforcement as a regulatory strategy, then, applies the structuring force of technology to the individual. Lessig argued that if citizens did not voice their preferences over the kinds of code that would be used to make up the Internet, that code would regulate behavior in ways that might be inconsistent with societal values. He posited that law might be used to shape technology (or code) in ways that are consistent with democratic principles, but he warned that code was so far being used to constrain behavior in ways that are not consistent with a democratic society.[11] This line of thinking can also benefit from the insights of science and technology studies.

The strategy of technological enforcement, as it is explored in the legal studies literature, is concerned primarily with technological enforcement's deterministic effects. It is related to the technology studies tradition that theorizes about how technologies come to structure the actions of individuals and societies. Technological enforcement is most prominently related to Langdon Winner's (1985) concept of the politics of technological artifacts. Winner's view is that much of the built/technological world either intentionally or unintentionally embodies power relations and worldviews that are consistent with the society and people who implement and design such technological structures. Therefore, technologies, through their use, subject the user to acting out those worldviews and power relations.[12] Because technologies linger throughout a society's history, they can continue to reproduce specific worldviews and ideologies invisibly over generations. This view is partially deterministic in its suggestion that society and individuals conform to technological structure imposed on them or that technology shapes society. The deterministic stance is countered by the understanding that society does indeed have a choice in technologies and a hand in its own technological regulation.

Winner's work goes hand in hand with the work of Richard Sclove (1995), whose central proposition is that society ought to make technological choices that are consistent with "strong democracy."[13] Sclove's call for democratic technologies suggests that the consequence of ill-conceived technological systems is the loss of democratic principles and institutions. Winner (1985) proposes a similar consequence of technological choice, noting, for example, that nuclear power necessitates a host of government and civil institutions to ensure its safe and secure use. Such institutions may necessarily infringe on privacy, increase secrecy in society, and have a whole set of unintended consequences for democratic institutions.

These issues are pertinent for the analysis of technological enforcement as an obstacle confronted by the digital rights movement because technologies that solidify positions in an ongoing legal debate (digital copy controls in the digital copyright debate) are potentially oppressive. Those who are not in a position to design technologies of their own or who are not in a position to participate in the policymaking process are effectively locked out of democracy. Furthermore, if technological enforcement is widely adopted, it becomes commonplace, and the behaviors that it regulates become more difficult to debate.

Responding to Technology—Resistance through Technology

This book also examines the use of hacking as technological resistance, a powerful extrainstitutional tactic for the digital rights movement. Technological resistance is a strategy wherein users/hackers design and deploy politically motivated technologies that challenge the digital copyright enforcement regime. Technological resistance is the opposition of technological enforcement by technological means expressly designed for such a purpose. Exploration of the technological enforcement/technological resistance dichotomy illustrates the regulatory force of technology, the role that government or other institutions may play in the design of technology, and how social movements can use the deterministic power of technology to counter regulatory attempts.

Technological resistance works against technological enforcement and the assumptions about behavior that technological enforcement embodies. It is the technomaterial expression of counterculture or counternorms. Technological resistance technologies[14] are used to counter technology-protection measures and the laws they espouse. By focusing on technological resistance, I deviate from what has become a dominant approach to understanding the role of technology in regulation. The majority of analysis on the subject has occupied itself with understanding the role of technology in regulation and how government has been using this strategy in digital copyright enforcement. In contrast, I approach the role of technology from the perspective of individuals who are trying to resist or subvert regulation with technology of their own. Technological resistance is the logical response to law embodied in and enforced with technology. The implications of this conclusion are potentially troubling because if technology is a powerful tool to resist unfair regulation, then only technologists have the know-how to exercise that power.[15]

Technological enforcement is an effective strategy for regulation because it has the power to settle ongoing debates about the balance of copyright, even though the balance of copyright ought to be an issue that is always debated and reexamined. On issues that ought to be always debated, should we want technology to enforce laws? Do we want to technologically close off issues that are continuously reshaped by courts? These questions are important to consider because the permanence of technology will make changes in the legal world more difficult to implement. As a consequence, the use of technological resistance will be a strategy for change that will become increasingly important in society.

User Agency and Technology

One last theme that runs through much of the discussion in this book is user agency. Although not something I discuss explicitly, the idea of user agency undergirds considerations of why users, their views on the use of the technology meant to mediate content matter, and their own conceptions of how technology should be used are factors that lawmakers and the content industry should consider. This theme is informed by research in technology studies that concerns itself with how society eventually comes to use technology. Informed by theories on the social construction of technology, the idea is that there is a period of negotiation among stakeholders in the design of a technology in which the meaning and use of the artifact are in flux, but after which the technological use and meaning become fixed. Responding to technological "closure," other research has pointed to the fact that many technologies and the systems they are embedded in are seldom completely "closed" to interpretation and appropriation and that users in fact continue to negotiate use and meaning long after a technology is release out in the "wild."

The question, then, becomes one of understanding users' motivation and the means by which they effectively resist closure. How do they come to exercise agency? In this book, users are considered a contested concept in the minds of activists, policymakers, and the content industry. All these players continuously define user identities, and all those identities are admittedly present within the panoply of consumers actually using and consuming media. The issue of user agency becomes important when a certain number of those users can effectively appropriate or redesign technology, which is then recycled into mass consumption.

This issue reflects the idea that users and technology are co-constructed. In the same fashion that, say, early computer hobbyists were constructed or imagined by designers and then assumed new roles as personal computers evolved, so, too, users of digital technologies were imagined but then assumed new roles, thus pushing digital technologies in new directions (Lindsay 2003). In the case of the digital rights movement, user agency is particularly powerful (if it were not, the content industry would not be spending billions in lawyers, lobbyists, and technology to limit it). Hackers and other less technologically savvy users are constantly seeking out ways to make existing technologies fit their personal expectations. Thus, as shown in later chapters, technologies such as hacks to eBook encryption and the iTunes digital rights management (DRM) system find a receptive user base among consumers who use these hacks to reclaim access or convenience in content consumption. In many cases, these technologies also allow for participation, which means they also may serve to construct uses: the user and the technologies of content consumption are cocreated as users discover ways of appropriating appropriation technologies.

This final point can be made clearer by considering the ways in which iTunes DRM system hacks themselves became reconfigured. Apple designed its DRM system to govern music consumption, conceptualizing a kind of music consumer in the process. Hackers who hacked the DRM system had their own visions of iTunes users (as consumers who would want to do more with the music than the iTunes end-user license agreement [EULA] would allow). Users themselves then did something else. Although the EULA was concerned with controlling the number of copies of a song, and the hacks to the DRM system undermined Apple's ability to enforce the EULA, some users didn't hack the DRM system simply so they could make more copies, but rather so that they could incorporate a song into a video they had made or sample the song for a DJing project. If these cases show us anything about user agency and technology, it is that technological meaning and functionality are open for interpretation and appropriation.

User agency also implies resistance to configurations of expected uses and to regulatory mechanisms. Because DRM enacts state policies, contracts, and copyright, it becomes important to see resistance through a technological lens and to configure technology not only as artifact, but as action, collective action.[16] Thus, technologies such as DRM straddle a number of important social domains—law, culture, and consumption. Technologies such as iTunes hacks likewise straddle law, culture, protest, and participation: they occupy those domains both physically and meaningfully. They are important beyond their function.

2 The National Information Infrastructure and the Policymaking Process

Who are the users of today's digital technologies? What are their expectations, and how are they effectively cocreated alongside the digital technologies that more and more are ubiquitously mediating culture? In an age where new technologies emerge into the consumer market at a blistering pace, it should strike no one as surprising that consumer expectations are created by the hype and realities of "ease of use," "portability," productivity, and connectivity. In other words, we are often sold gizmos and gadgetry imagined for us but not yet defined by us and our unique uses. So when technologies of this sort come into our possession, they morph into something sometimes unimagined by their designers. They are often broken into, hacked, glitched, or worked around. More and more media corporations and technology makers have come to fear these activities even as they ironically strive to sell technology as something uniquely personal, tailor made for me (or you) alone. Technology companies hope to control appropriations and to limit user control of technologies and content contained therein at the same time that their rhetoric (and, to be fair, some of their design) seeks to convince us that we are at the helm, plotting the course of use. As Tarleton Gillespie (2007) has shown us, designers often effectively "frustrate" certain uses in order to preserve copyright or licensing controls over content. However, the history of technology shows that these efforts have sometimes been met with user resistance and that there is often a degree of plasticity in the ways technologies are adopted and used (Oudshoorn and Pinch 2003).

Laws and lawmakers, for their part, struggle with user forays into agency and appropriation, often ignoring or downplaying legitimate user claims about their right to control the kinds of uses they want to make. In the case of the early days of the digital rights movement in the mid-1990s, the struggle over how to reign in unforeseen consequences of digital technology became centered on Internet policy. Rightly seeing the Web as an extremely convenient way to distribute content and as the key technology that would connect consumer-owned media (CDs, DVDs, eBooks, etc.) to easily accessible distribution systems that at the time could only be imagined (but that later became Napster, Grokster, torrent technology, etc.), content owners began to fret and lobbied lawmakers for more stringent technolegal protections. And so it came to be that the early battleground for the digital rights movement centered on copyright law, the legal instrument (along with licensing) that has traditionally regulated and protected cultural products conveyed via media (digital or otherwise).

I purposefully focus here on the legislative history of one law affecting copyright in particular, the DMCA, because its deliberations, the debates that are part of the Congressional Record, and its ultimate enactment illustrate the issues that confronted policymakers, content owners, and some consumer representatives. Deliberation for policymakers began with the call to develop guidelines to regulate the National Information Infrastructure (NII), what is now known as the Internet or World Wide Web. The DMCA itself was a result of these deliberations as well as a desire by lawmakers and copyright holders to normalize US copyright statute with emerging global intellectual-property law regimes. Review of these documents illuminates the way legislators eventually came to see the challenges of digital technology for the cultural industries and points out some of the blind spots in those views that eventually both shaped law and spurred activism against it. Subsequent sections address the final recommendation for the DMCA and the relevant sections of the DMCA.

A very short summary of the US copyright statute is prudent first because, as will become apparent, although much of the early activism on digital rights targeted the DMCA, it happened in the shadow of copyright law and its rationale.

A Brief History of Copyright in the United States

The Constitution grants Congress the power "to promote the progress of science and useful arts, by securing for limited times to authors and inventors the exclusive right to their respective writings and discoveries" (Art 1, sec. 8, clause 8; Copyright Clause). In response to this mandate, Congress has developed a series of laws that protect intellectual property in a variety of forms. For example, an individual with a unique idea may choose to patent his or her invention, thus gaining a temporary legal monopoly over the use of that invention as well as a stake in any subsequent inventions derived from the protected work (in the form of royalties and permissions).

Copyright law grants creators of written works the sole right to reproduce and perform their works, and trademark law protects brand marks from reproductions that may undermine the market reputation of that mark's owner.

Congress first codified copyright law in Title 17 of the US Code of Law in 1790. Since its enactment, the copyright statute has gone through four major revisions. The Copyright Act of 1790 applied the Constitution's intellectual-property provisions to copyright. It granted American authors a limited monopoly over their work and gave them the exclusive right to print and reprint their works for fourteen years. At the end of the fourteen-year monopoly, authors had the option to extend their protection for another fourteen years. Thus, per the 1790 Copyright Act, an author had exclusive copyright over his or her work for a total of twenty-eight years.

In 1831, Congress revised the Copyright Act for the first time since its enactment, extending the initial time limitation on exclusive copyrights from fourteen to twenty-eight years, with an option to extend it for an additional fourteen years. As a result, by 1831 authors could hold copyright over their works for a total of forty-two years. With the 1870 revision of the act, Congress shifted the administration of copyright from district courts to the Library of Congress Copyright Office but did not extend the term limits.

Today, the 1909 Copyright Act is considered a major revision of the copyright statute because it expanded the categories of copyrightable material beyond literary works to all works of authorship (Litman 2001). In addition, Congress again extended the copyright term, raising the potential number of years an author could hold copyright to fifty-six years. A copyright holder now had twenty-eight years of protection with the possibility of further extension of the application for an additional twenty-eight years.

The most recent major revision to the copyright statute is the Copyright Act of 1976. In the 215-year history of federal copyright law, the 1976 act is the most expansive revision, enacted in part as a direct response to the emergence of new technologies that could affect a copyright owner's ability to exercise his or her rights. Although the act was revised in 1909 in part to respond to the developments in technology, such as the piano roll and the talking machine, adjusting to technological change was not its primary aim. Rather, it was meant to normalize copyright law, which had become an amalgamation of the various major and minor revisions since 1790 (Litman 2001). In comparison, the 1976 revision extended term limits of copyright to the life of the author plus fifty years and extended the term limit of works for hire (works owned by someone other than the author) to seventy-five years.[1] The 1976 revision covered a great deal of ground: defining the scope of copyright, codifying the fair-use and first-sale doctrines, defining copyrightable material, and defining remedies against copyright violation. Since the enactment of this revision, the copyright statute has received a host of amendments and expansions, of which the DMCA is but one.

The NII

Shortly after the 1992 elections, the Clinton administration appointed Secretary of Commerce Ronald H. Brown to head the newly convened Information Infrastructure Task Force (IITF) and charged it with formulating telecommunications and information policies that represented the administration's vision of the NII (Litman 2001).[2] The IITF envisioned the NII as "a seamless web of communications networks, computers, databases, and consumer electronics that will put vast amounts of information at users' fingertips" (Brown 1992). The IITF noted the primacy of businesses in building this communications and consumption infrastructure and, as a result, wanted government to facilitate the work that businesses were doing. To achieve this goal, the IITF's policy recommendations were guided by objectives such as providing tax incentives to business, helping businesses develop technologies, ensuring protection of intellectual-property rights and recognizing the NII's potential for being borderless, and coordinating with other governments for the regulation of a global information infrastructure (Brown 1992).

The IITF organized itself into three committees: "[t]he Telecommunications Policy Committee, which formulated Administration positions on key telecommunications issues; the Committee on Applications and Technology, which coordinated Administration efforts to develop, demonstrate and promote applications of information technologies in key areas; and the Information Policy Committee, which addressed critical information policy issues that must be dealt with if the NII is to be fully deployed and utilized" (Lehman 1994). The Working Group on Intellectual Property Rights (WGIP), chaired by Assistant Secretary of Commerce and Commissioner of Patents and Trademarks Bruce A. Lehman, was a subcommittee of the Information Policy Committee. Its objectives mirrored those of the IITF, and it sought to achieve those objectives by analyzing how copyright might have to change to protect content on the NII. In July 1994, the WGIP released the first draft of its recommendations for amendments to the US copyright statute based on the technological implications of the NII. The Green Paper (WGIP 1994), as the preliminary report came to be called, was written in the shadow of the copyright statute and directly addressed how content in digital networks and embedded in digital media would be protected.

Prior to the release of the Green Paper, the WGIP convened hearings in Washington, DC, Los Angeles, and Chicago. The hearings were meant to allow interested parties the opportunity to provide suggestions to the WGIP regarding the kind of policy needed to protect copyright on the NII. These testimonies and written comments submitted beforehand informed the Green Paper and give an important sense of how representatives of the cultural industries as well as some consumer representatives saw the challenges of the emerging NII.

The WGIP posed the following questions to respondents in the notice for the hearings:

1. Is the existing copyright law adequate to protect the rights of those who will make their work available via the NII?
2. Do the existing fair-use provisions of copyright law adequately accommodate the interests of the users of the works available via the NII? What statutory or regulatory changes, if any, should be made?
3. Should standards or other requirements be adopted for the labeling or encoding of works available via the NII so that copyright owners and users can identify copyrighted works and the conditions for their use?
4. Should standards be established to encourage or require intercommunications or exchange of information and the interoperability of the different types of computer software and systems supporting or utilizing the NII?
5. Should a licensing system be developed for certain uses of any or all works available via the NII? If so, should there be a single type of licensing, or should the NII support a multiplicity of licensing systems?
6. Are there technical means for preventing unauthorized reproduction or other unauthorized uses of copyrighted works that should be mandated or required to comply with certain standards (similar to the serial-copying controls required in digital audio-recording devices and digital audio-interface devices under the Audio Home Recording Act of 1992)?
7. What types of educational programs might be developed to increase public awareness of intellectual-property laws, their importance to the economy, and their application to works available on the NII? (*Federal Register* 58-[53917])

Imagining the Web: A Clear Vision of an Unknown Future

In response to this call for policy suggestions, eighty-two written comments from seventy-two organizations and individuals were submitted to the WGIP prior to the hearing, addressing the various questions it had posed.[3] Of the written comments submitted for the record, twice as many were from representatives of the software, publishing, motion picture, and music industries as those submitted on behalf of user or consumer interests. Furthermore, whereas organizations such as the Home Recording Rights Coalition and the Electronics Industry Association nominally represented consumers, none represented media users outside of institutional frameworks (such as libraries) or outside of the "consumer" category (such as customers for the VCR industry).

At the hearings, organizations and individuals gave testimony reiterating the position they presented in their written comments. Predominantly present at the hearing were copyright owners. Of the twenty-five or more testimonies at the hearings, only six came from libraries and universities, and only one came from the Electronics Industry Association, representing consumer use of potentially infringing electronics (see table 2.1).

When asked if existing copyright law was adequate for protecting copyrighted material over the NII, content owners overwhelmingly said it was. Fritz Attaway, representing the Motion Picture Association of America (MPAA), noted that "[w]ith relatively modest fine-tuning, including the recognition of performance rights for sound recordings, our existing copyright law is adequate to protect the rights of those who will make their works available via the NII. . . . [I]f existing law is given broad application to protect the rights of copyright owners, few statutory or regulatory changes should be necessary" (in *Comments* 1993).

Copyright owners' major concern was over how existing definitions of distribution, transmission, and copying would be enforced in the NII. In this regard, Mark Traphagen, representing the Association of Software Publishers, commented that "[c]opyright and other intellectual property rights must be respected regardless of the technological means by which they are presented or disseminated for rights holders to make their works available at all. Therefore, it should be made clear that the exclusive rights provided to copyright owners, as well as other intellectual property rights, must be respected on the NII" (in *Comments* 1993).

Content owners, in affirming the adequacy of the copyright statute for the NII, portrayed fair use in a way that constricted personal uses and proposed that the statute and the courts already agreed with their viewpoint. Libraries and other public institutions, in contrast, argued for the preservation of fair use in terms of archiving and other processes that would ensure their continued operation in the digital domain but did not push the notion that the consumer might have other personal interests or that the emergence of the NII and digital technology might constitute a reason to expand fair use and give users facilitated access to cultural products. For example, Robert Oakley, the director of the Law Library at Georgetown Law School, understood libraries to be the primary mediators of fair use and argued for the preservation of their exemptions. He noted that "the fundamental purpose of copyright is for the public good [and] to achieve the goal of promoting the public good. . . . [T]he NII should preserve fair use in the library exemptions and allow for a variety of pricing structures" (in *Testimony* 1994).

Some individuals responded to issues concerning potential legal and technological limits to fair use and personal use by suggesting that a licensing scheme, dictated by the market, would be best suited to structure the types of uses that would be legal for purchasers of copyrighted content. For example, Steven Metalitz, speaking for the Information Industry Association, noted,

> Marketplace trends have also influenced the development of fair use concepts. In particular, the growing trend toward defining permissible uses of copyrighted material by contract is a positive development for both copyright owners and users of copyrighted material. Technological changes accompanying NII development could reinforce this trend and should therefore be encouraged. . . . In practice . . . we often look to a contract to specify the degree to which (and the circumstances under which) the author will permit another to exercise the exclusive right created by copyright law. Without the ability to license exercise of exclusive rights through contract, a marketplace for digital information products would be severely constrained. (in *Comments* 1993)

This comment confuses the origins of the exclusive rights granted to authors by copyright and the limits of those rights outlined in the statute under the fair-use doctrine. It implies that contracts ought to govern fair use when in fact the copyright statute and court doctrine delineate what is and is not fair use. Metalitz's view allows content owners to be the arbiters of boundaries of fair use by implementing contract and skirting delineations in the statute and case law. The suggestion that contract ought to dictate the conditions of fair use put individual users at a great disadvantage. The only users in a significant position to bargain for rights of use are large institutional consumers who can act as gateways for large groups of users.

Last, in response to questions about technological measures to enforce copyright industry interests, commenters concurred that technological protection measures would be vital and that laws to protect those measures would be necessary. Metalitz emphasized that it would be "appropriate . . .

Table 2.1
List of Organizations Present at the IITF WGIP Public Hearings at Andrew Mellon Auditorium, Washington, DC, September 1994

Witness	Position	Organization
Steven J. Metalitz	vice president (VP) and general counsel	Information Industry Association
Maria Pallante	executive director	National Writers' Union
Stephen Haynes	manager	West Publishing Company
Lisa Freeman	director and chair	Electronic Caucus Association of American University Presses
Timothy King	VP planning and development	John Wiley Publishers
Robert Oakley	director of law library	numerous library and education associations
Joseph Cosgrove	professor	Department of Political Science, King's College
Dennis Bybee	not listed	International Society for Technology in Education
David Rothman	private individual	none
David Pierce	president	American Association of Colleges
Fritz Attaway	senior VP	MPAA
Bernard Sorkin	VP and general counsel	Time Warner
Hilary Rosen	VP	RIAA
Lawrence Kenswill	VP	MCA Music Entertainment Group
Richard Ducey	VP	National Association of Broadcasters Research and Information Group

Table 2.1
(continued)

Witness	Position	Organization
Benjamin Ivins	general counsel	National Association of Broadcasters Research and Information Group
John Masten	VP	New York Public Library
Gary Griswold	president	Infologic Software
Robert Kahn	president	National Research Initiatives
Brad Cox	not listed	Center Electronics Markets
Ronald Laurie	not listed	Weil Gotshal and Manges
Henry Perritt	assistant law professor	Villanova University
Ronald Palenski	VP	Information Technology Association of America
Mark Traphagen and Irene Rosenthal	not listed	Software Publishers Association
Thomas Lemberg	VP	Lotus Development Corp., on behalf of Business Software Alliance and Alliance to Promote Software Innovation
Brian Kahin and Philip Dodds	not listed	Interactive Multimedia Association
Gary Shapiro	VP	Consumer Electronics Group Electronic Industries Association
Douglas Brotz	not listed	Adobe Systems
Frank Connoly	assistant professor	American University

to consider legal sanctions against the distribution of devices or techniques whose primary use is to defeat or circumvent intellectual property management technology" (in *Comments* 1993).

Review of the proceedings of these hearings suggests that the use of technological protection measures to enforce copyright on the Internet was a foregone conclusion on the part of the WGIP. The fact that the WGIP did not ask whether these measures were desirable but rather wondered whether there ought to be a government mandate for standardization among them is quite telling. Technological enforcement had precedent in the form of the serial-copy controls of the Audio Home Recording Act (AHRA) of 1992. As far as the WGIP was concerned, that mode of enforcement was a legitimate way to curb unwanted uses of copyrighted works.[4] The WGIP never raised questions regarding the technology's unintended consequences, its long-term effects on fair use and personal uses, and the implications of the adoption of such a technolegal regime for the constitutional intent of intellectual property and, more broadly, democracy. Thus, the blind spots in this policy process lay not only in inadequate reflection on personal noncommercial uses, but in the very questions posed by WGIP.

Industry positions can be summarized succinctly: industry representatives wanted to extend copyright to the NII and asked for no sweeping changes to the statute; they framed fair use as being properly understood in the law and felt it needed no further protections; and they thought that technological measures to protect and enforce copyright (along with laws to supplement them) were necessary. The discussion of fair use on their part gave a reading that was least problematic from their standpoint. They framed their views as a defense against charges of infringement rather than rights and left untreated the arguments that might frame fair use as a right, especially when considered in light of its importance for free speech.

What is analytically important is not so much that industry representatives responded in this fashion, but rather that the prompts from the WGIP didn't ask them to think beyond the constraints of copyright. The prompts, then, served as a kind of template, presupposing the status quo in copyright protection or suggesting its extension. The first prompt, which asked respondents if existing copyright was adequate for protecting copyrighted material over the NII, is a good example. The assumption was implicit that the NII would contain copyrighted material. Of course it would, but the notion that copyright, as it was, needed to be extended to the NII was posed as a given. Furthermore, the idea that the NII (and ultimately the World Wide Web) would be a medium against which copyright law would necessarily be applied in existing or extended fashion glosses over other possibilities: that maybe the technological realities, the potential user practices, and the benefits to free speech or creativity would require less copyright protection or a different kind of protection on the Web.

To be fair, policymakers did not assume that the Web or digital technologies would turn out to be like any other medium; if that had been their assumption, then the hearings would not have been necessary. Laws would have simply applied easily to the new technological regime of creation and distribution. The working group and policymakers at the time recognized the challenges digital technology would pose, but their approach was conservative. In other words, the first prompt really was asking, "How can we preserve the current level of protection for cultural products?" This critique does not suggest that lawmakers should have had a better vision of what would happen on the Web or to digital technologies with regard to user practices, but rather that the process of thinking about the technology and cultural production as well as the application of law to those things were entrenched in a particular worldview about the overall nature of cultural production—that valuable creative works would come from primarily established cultural industries. When the WGIP's prompt asked about "those who will make works available via the NII," the group was thinking of institutional cultural industries.

In the zealous endorsement for extending and expanding copyright to the NII, some important realities that were not part of content control in the analog world but that became important in the digital world were ignored. First, technological realities in media created a gap between the letter of the law and what was technically enforceable. Prior to the emergence of the digital technologies related to the NII, for example, personal noncommercial infringements on copyright owners' rights that fell outside of what content owners considered fair use were unpreventable. The copyright owner of a song could not prevent a person from making multiple copies of an audiocassette recording and distributing it to as many of her friends as she desired.

For their part, copyright owners relied on the technological inconvenience of copying and distributing analog media to limit the extent of unauthorized personal uses, although always considering unauthorized copying and distribution an infringement that ought to come under control. Consumers have for some time thought differently, with sampling and "mix tapes" being a typical part of some music fans and artists' experience. Content owners thought copyright needed reworking in the NII, and the fact that digital technology allowed for easy copying and distribution of content was perceived as a threat. This threat became a rallying call for content owners to close the gap between copyright law and personal uses. The preferred method for closing this gap was the implementation of access and copy-control technology along with anticircumvention laws to keep people from breaking these technological controls. The implementation of these laws and control technology amounted to more than a "minor change." By creating technological and legal structures that would regulate the breadth of access a consumer may have to digital media, the law would in fact have a significant impact on consumer behavior.

In testimony, the matter of consumer custom and user engagement with media was not thoughtfully considered. When the industry spoke of users, it spoke of them at best as consumers, noting that value in the NII would be generated along the same lines as it had been before: content would be made by the industry, and people would buy it. When Internet and digital media users were explicitly imagined, they often were thought of as pirates or potential pirates who would need to be educated on the rights of authors or to be technologically protected against.

The consequences of technologically protecting against the threat of personal uses and its impact on legal privileges (such as fair use) were not properly considered. Technological protection measures that would close off personal uses would also close off legally defined fair uses because protection technologies have no way of differentiating between fair or infringing access. No one discussed how this unintended consequence would be addressed. The DMCA in its final form included a weak technological solution for this problem, prompting legal scholar Pamela Samuelson (1999) to ask whether Congress intended to give the public hollow privileges over digital content.

The Green Paper and the 1976 Copyright Act

The Green Paper (WGIP 1994), a result of the meetings, testimonies, and comments described earlier, was the WGIP's first attempt at formulating a policy that would address issues of copyright in the NII as well as international intellectual-property treaty obligations for the United States. As noted earlier, this review of the documents and legislative history shows how policymakers and content owners saw the challenges posed by emerging digital media in the 1990s. The way they saw these challenges (as well as opportunities) colored how they proposed policy change. The Green Paper addressed a number of issues in its recommendations, but for our purposes and for the purposes of understanding what laws first mobilized digital rights activists, the recommendations for transmission, first sale, fair use, and technology are discussed here.

In its review of the various provisions of the Copyright Act, the WGIP pointed out the effect of emerging information technology on the copyright statute (Lehman 1994).[5] Based primarily on cultural industry recommendations, the Green Paper set out to preserve copyright protections in the NII; however, because of the technological realities of digital media (explained more fully later in this chapter), that task would require some important redefinition of key terms in the statute to adapt copyright to digital technology. The Green Paper's recommendations can be classified as follows: (1) proposals for law, (2) proposals for technology, and (3) proposals for education.

Proposals for Law: Transmission and Making Copies

In its policy recommendations, the Green Paper first addressed the issue of transmissions of content over the NII because the working group felt the NII's impact on definitions of transmission would be important.

Authors' rights over their works are outlined in section 106 of the US Copyright Act of 1976.[6] They include the right

(1) to reproduce the copyrighted work in copies or phonorecords;
(2) to prepare derivative works based upon the copyrighted work;
(3) to distribute copies or phonorecords of the copyrighted work to the public by sale or other transfer of ownership, or by rental, lease, or lending;
(4) in the case of literary, musical, dramatic, and choreographic works, pantomimes, and motion pictures and other audiovisual works, to perform the copyrighted work publicly;
(5) in the case of literary, musical, dramatic, and choreographic works, pantomimes, and pictorial, graphic, or sculptural works, including the individual images of a motion picture or other audiovisual work, to display the copyrighted work publicly; and
(6) in the case of sound recordings, to perform the copyrighted work publicly by means of a digital audio transmission.

Important elements of these "fundamental" rights are the terms *copies* and *phonorecords* as well as their definitions.[7] *The definitions assume that copies and phonorecords are tangible objects.* Both of these definitions therefore imply that *fixation* of content is a central component of what constitutes a legitimate copy (the generic term *copy* is used to refer to both copies and phonorecords as defined by the Copyright Act). The WGIP thought that the key to preserving authors' rights in the NII would be to maintain the applicability of definitions in the statute. A central concern for why this definition might be in jeopardy was the possibility that businesses or individuals may distribute copies of content over the NII and that the manner of conveyance might be construed as a "transmission" (a process an author does not necessarily have a right to control) rather than a "distribution" (a process an author does have the right to control). The Green Paper pointed out that a transmission is not a fixation and as such is not protected by the Copyright Act in the same manner as copies unless a copy is simultaneously made.[8]

The WGIP feared that conveyance of content via the NII, because of the manner of conveyance, might be construed as a transmission without fixation, despite the Copyright Act's legislative history, which more or less inoculates against that interpretation of "transmission." In such a case, the transmission would not fall under the protection of the Copyright Act.[9] When the Green Paper contemplated the consequences for the legal notions of transmissions and copies, it framed them with the goal of preventing transmissions via the NII from being considered ephemeral. Beyond wanting transmission classified as a method of distribution, the WGIP also wanted the term *transmission* to encompass all types of content communication via the NII. As such, the WGIP suggested transmission would not only be a form of distribution of copies, but also a form of communicating public performances and displays as well as the making of a copy. This reformulation of transmission is expansive to say the least and stretches the term to its conceptual limits. It would conflate processes in the conveyance of content that had once been clearly distinct.

Proposals for Law: First Sale

Next, the Green Paper addressed the issue of first sale.[10] Noting that if electronic transmissions were viewed as distributions of copies, it indicated that those distributions would be subject to the first-sale doctrine. The Green Paper foresaw a problem if digital content was further distributed—in the same manner that physical books can be resold or donated—by people who had legally purchased that content. This prospect was and continues to be frightening for copyright owners because digital distribution would necessarily make a copy of the work while in the process of conveyance, and there would be no guarantee that the original copy would be deleted after the distribution.[11] To address this issue, the Green Paper suggested that in transactions of content via the NII, both the distribution right (right number 3 in the list of rights from section 106 of the 1976 Copyright Act) and the right to make copies are implicated. The Green Paper expanded its definition of distribution of content on the NII to include not only transmission, but also the making of copies. The WGIP then suggested that the fundamental right would preclude further distribution of digital copies over the NII via the first-sale privilege because such distribution would be a violation of the copyright owner's exclusive right to make copies. By melding the definitions of the terms *distribution, transmission,* and *making copies,* the WGIP placed digital content within a hierarchy of rights and privileges in copyright, where the right to make copies afforded to authors by the statute precludes the first-sale privilege.

The WGIP recommended that first sale not be extended to copies of content acquired through distribution by transmission. The Green Paper clearly wanted the copyright statute to treat transmissions as distributions, copying, and/or public performances/displays so that it could grant authors rights to their work. At the same time, in cases where the consumer was ready to benefit from first sale, the WGIP wanted the law to ignore the fact that publishers were indeed *distributing* their goods because that first sale is specifically a limitation on the distribution right. If transmission via the NII were considered solely a distribution, then the consumer would be able to exercise first sale. The WGIP wanted the "fundamental author rights" (the rights to make and distribute copies) to be central in interpreting cases where copies of content were distributed without the copyright owner's consent because this approach would preclude the applicability of the first-sale privilege.

The recommendation for the redefinition of *transmission* would essentially explode its meaning, so that a transmission in legal terms would mean all possible acts for which a content creator has an exclusive right—distribution and the making of copies being the most important. The transmission of copies would be redefined as a type of distribution and therefore be governed by the right to distribute copies when copyright owners were concerned, and any further transmission by consumers exercising the first-sale privilege would be governed by the "fundamental rights." The recommendations regarding first sale, then, created an exemption from first-sale privileges for copies of content acquired through transmission.

The heart of this problem lies beyond semantics. At the core is the NII's effect on legal definitions of transmission, distribution, and the making of copies. What the Green Paper does not overtly mention is that transmission, distribution, and the making of copies are all one in the same in the context of the NII. It implicitly recognizes this view by recommending that the definition of *transmission* be expanded, but this recommendation leaves one with a feeling of unfairness to the consumer. Because the method of distribution changed, the WGIP expanded legal definitions to stem unwanted circumstances for copyright owners. Yet such an expansive redefinition occurred at the expense of consumer privileges.

The Green Paper lacked analysis in its attempt to rationalize exempting digital copies from first sale. Although it is clear that copyright owners are potentially harmed from the distribution of digital media if original copies remain in the distributing systems, there is no analysis or mention of the market benefits of reselling digital goods. After passage of the DMCA (based on the Green Paper and the revision of it in the final draft, the White Paper [WGIP 1995]), however, Congress mandated that the US Copyright Office evaluate the effects of the DMCA on first sale. In sum, the registrar of copyrights recommended against creating the "digital first-sale" privilege, a decision based primarily on copyright owners' arguments of potential market harm (US Copyright Office 2001). It seems that fear of the unknown in this respect preempted exploration of new markets.

For the WGIP, preserving publisher rights for content on the NII was the central goal and redefining key legal terms was one way in which to achieve it. The WGIP, when faced with the NII's possibilities, found itself with a problem of metaphysical proportions. To ensure a copyright owner's right in controlling copies and phonorecords, it pointed out that transmission is a form of immediate fixation in which copies are distributed, implying that the distribution right and the right to make copies apply. It also noted that because many people can potentially view a single transmission, that transmission, under certain circumstances, might constitute a public performance and display. However, to ensure continued control over that fixed media and to exploit the low cost of distribution, the Green Paper made the applicability of the distribution right serve only copyright owners during initial distribution and focused on the right to make copies when it became apparent that consumers may want to further distribute those copies.

The definitions of *transmission*, *distribution*, *copying*, and *public displays/performances* are strained by the technologies of the NII (computers and the Internet, for example) because they have abstracted content from their tangible confines. Distribution, copying, public performances/displays, and transmission, previously separate processes, have converged into a hybrid process made possible by digital technologies. Rather than reevaluate the hybrid processes anew with respect to implications for both distributors and consumers, the Green Paper fit old definitions into new contexts, making them benefit only the distributors of content.

Proposals for Law: Fair Use

The fair-use doctrine has historically been defined in the courts and was codified in the 1976 Copyright Act. Section 107 of the act states:

> Notwithstanding the provisions of sections 106 and 106A, the fair use of a copyrighted work, including such use by reproduction in copies or phonorecords or by any other means specified by that section, for purposes such as criticism, comment, news reporting, teaching (including multiple copies for classroom use), scholarship, or research, is not an infringement of copyright. In determining whether the use made of a work in any particular case is a fair use the factors to be considered shall include—
>
> (1) the purpose and character of the use, including whether such use is of a commercial nature or is for nonprofit educational purposes; (2) the nature of the copyrighted work; (3) the amount and substantiality of the portion used in relation to the copyrighted work as a whole; and (4) the effect of the use upon the potential market for or value of the copyrighted work.

The issue of fair use is central to many of the debates that surround technology and copyright of digital content. The popular and legal meanings of *fair use* are in fact at the heart of what the digital rights movement would like to capture and are discussed further in later chapters. For now, it should suffice to say that the WGIP's Green Paper argued that fair-use case law supported claims by copyright owners (in their written comments) that fair use is a highly regulated doctrine relying heavily on the fourth factor listed, "the effect of the use upon the potential market for or value of the copyrighted work." The Green Paper argued that the burden of proof is on the potential infringers to show that their use was noninfringing.[12] The Green Paper noted that so long as the courts continued to interpret law as they had done in the past, clear-cut cases of fair use and infringement on the NII would be properly decided.

The Green Paper mentioned two cases that support its claim that courts had already applied proper guidelines to questions of fair use on the NII. The first was Playboy Enterprises v. Frena (839 1552 [1993]) in which courts ruled that uploading and downloading proprietary images by users on George Frena's electronic bulletin board system did not constitute a fair use because of the effect that such transmission of content could have on the market for *Playboy*'s images. As a consequence, the court held Frena guilty of contributory infringement of copyright. The second case was Sega Enterprise Ltd. v. MAPHIA (948 F.Supp. 923, 41 USPQ2d 1705 [1996]), involving another electronic bulletin board service. Much like in the *Playboy* case, the courts found that MAPHIA was contributing to copyright infringement by its users when they uploaded copies of Sega's videogames. The court argued that such uploading could have a detrimental impact on Sega's market. In its recommendations, the WGIP did note that there might be some adverse effects to fair use as a result of its call for technolegal copyright-protection measures and so it convened a Conference on Fair Use to discuss these potential impacts (Lehman 1994).[13]

The Green Paper failed to take into account the meaning of fair use as defined by common user practices and anticipated that users would simply acquiesce to the dictates of the technological regimes designed by copyright owners. In this respect, the WGIP was short-sighted or chose to ignore the implications of a case from ten years earlier, Sony Corp of America. v. Universal City Studies Inc. (464 US 417 [1984]), which it reviewed only in passing, thinking that it lent support to its claims that fair use is strictly defined and contingent on the absence of market harm to copyright owners. In truth, the Sony case was unique in that it defined copying of copyrighted material on a VCR tape for the purpose of "time shifting" as a fair use and refused to hold Sony responsible for contributory infringement for its manufacture of the VCR. This court decision opened up the door for the manufacture and design of copying technologies with potentially infringing uses so long as there is no demonstrable market harm and that the primary uses of the technology are not infringing.

The Green Paper underestimated user resourcefulness and determination. Some consumers had a significantly different reading of the meaning of fair use from the one articulated by the Green Paper and in practice would adopt user-centered notions of fair use and interpretations of the Sony case as defenses in the design and distribution of peer-to-peer technologies, the design of circumvention technologies to access content on DVDs, and the design of technologies to circumvent DRM systems.

In sum, the WGIP thought that the NII would implicate fair use but that courts would continue to have a strict reading of what constitutes fair use. Keeping its recommendations for technolegal protection measures in mind, the WGIP noted that such measures might excessively expand authors' rights to control their work and recommended a Conference on Fair Use to discuss those effects.[14] The WIPG chose to ignore users' common practices concerning media and popular interpretations of fair use. As a result, it was blind to the potential legal storm that was brewing over the use and design of circumvention technologies that allow users to access or copy technologically protected media.

Proposals for Law: Anticircumvention Provisions

The Green Paper gave clear recommendations concerning the role of technology in ensuring copyright owners' interests. Its recommendations included suggestions for outlawing technological measures that may circumvent content-protection technologies and are the core of the DMCA section that has inspired backlash from activists. Technology is the entry point for digital content; such content cannot be accessed without it. As such, the WGIP saw technology as a central means of guarding against the negative consequences that the NII would have for content owners' rights. It follows that these technologies ought to be protected by law. In other words, copy-protection and access-control technologies are enforcement technologies for the provisions of the Copyright Act that give an author his or her rights. The DMCA's anticircumvention provisions (based on the WGIP's recommendations) are laws that protect those technologies. The WGIP implicitly understood technological protection measures as the best way to protect copyright owners' rights on the NII because of the difficulty in tracking infringers. The WGIP proposed thorough technolegal protection of content as a means of incentivizing for copyright owners to participate and contribute to the NII (Lehman 1994).

Laws that protect content by outlawing technologies and conduct that circumvent copyright protection are not without precedent. The Green Paper noted that the AHRA of 1992 had already done this for the serial-copy-management system on digital tape recorders and that the Telecommunications Act of 1996[15] protected encryption of satellite cable programming. These laws and others account for a trend in legislation that protects copyrighted digital content. The convergence of law and technology ensures protection, while at the same time outlawing technologies that would challenge those technolegal systems.

The Green Paper was relatively heavy-handed in allowing copyright owners protection against circumvention technologies, in proposing broad amendments to the Copyright Act, and in avoiding thorough discussion of how those proposed amendments would affect fair use, first sale, and technological innovation. The amendments include the following:

1. The addition of section 512 to read: "No person shall import, manufacture or distribute any device, product, or component incorporated into a device or product, or offer or perform any service, the primary purpose or effect of which is to avoid, bypass, remove, deactivate, or otherwise circumvent, without authority of the copyright owner or the law, any process, treatment, mechanism or system which prevents or inhibits the exercise of any of the exclusive rights under Section 106."

2. An amendment for section 501 defining an infringer as "anyone who violates section 512 is an infringer of the copyright in a work that utilizes the process, treatment, mechanism or system which the violator's device, product, component or service circumvents."

3. Amendments to section 503 granting courts powers in civil cases where section 512 had potentially been violated, noting,

> At any time while an action under this title is pending, the court may order the impounding . . . and of all devices, products or components claimed to have been imported, manufactured or distributed in violation of section 512.
>
> As part of a final judgment or decree, the court may order the destruction of all devices, products or components found to have been imported, manufactured or distributed in violation of section 512.

The Green Paper did not thoughtfully address technologies that might be used to access technologically protected works whose copyright had expired or a technology that would allow access to works for the purposes of fair use. For example, in what appears to be an overly simplistic explanation of why technologies designed to access works whose copyright had expired would not constitute infringement, the Green Paper noted that "the 'primary purpose or effect' standard will allow for the distribution of devices that deactivate the anti-copying systems used in such works, and that the benefits of the proposed legislation outweigh the possible problems" (quoted in Lehman 1994). The "primary purpose standard" would dictate that the main function of a circumvention technology is to circumvent protection technology on works that are no longer covered by copyright. The standard ignores the fact, however, that copy-protection technologies might be similar whether they protect a work still under copyright or not. Thus, it is impossible to distinguish the primary purpose of a circumvention technology as the Green Paper proposes because such technology would target copy protection generally and would not be specific to one instance in which the work protected had lost copyright.

Another argument against the Green Paper's anticircumvention provisions was that laws that outlaw technologies stymie technological innovation. Opponents of this argument have suggested that the Universal v. Sony case had established a doctrine that would still allow technological development of potential infringing technologies if they have some other significant commercial purpose.[16] The problem with this view is that it ignores the frailty of emerging technologies in a hostile legal environment; the process of establishing alternative valid uses is inherently contentious, and developers of technologies are almost always at a disadvantage when called on in the courtroom to qualify the other legitimate uses of technologies.[17] Furthermore, if the "noninfringing use" of a technology is defined in part by the market, many technologies designed with little commercial intent in mind are left exposed to regulation.

Perhaps most important were the tones of alarm expressed by representatives of the cultural industry as they reflected on the possible rampant copyright violation that would ensue on the NII if intellectual-property products were not properly protected. In those instances when industry representatives imagined threats from users of the NII, they did so to paint a picture of copyright pirates. Steven Metalitz from the Information Industry Association noted:

> While the new information infrastructure offers unparalleled opportunities for the widespread dissemination of this intellectual property to authorized users, it also offers unparalleled threats to [the] exercise of the exclusive rights of authors to reproduce, display and adapt these works. The same capabilities that give advanced information infrastructure its awesome potential also invite an epidemic of abuse of intellectual property rights. If we do not prevent that epidemic our hopes for what the NII can deliver to our workplaces, schools and homes will be blighted. (in *Comments* 1993)

Although there was of course some cause for concern (Napster showed the excesses of the type of infringement possible), the discourse of alarm set the tone and eclipsed the possibility of more reasonable discussion of how business models might function on the Web so that consumers might opt for purchasing content rather than downloading it for free or how business systems might be remodeled to allow for consumers' access and ease. Online digital-content-distribution models that are now wildly successful were a consequence of consumer behavior that could have been predicted and constructively addressed rather than framed as an epidemic.

The release of the Green Paper was received with a mixture of criticism and support. Following its release, the WGIP held a series of hearings in Washington, DC, Chicago, and Los Angeles and solicited and received written comments, which it then incorporated into the second version of the policy proposal, called the "White Paper" (WGIF 1995).

Conclusion

With respect to the digital rights movement, the Green paper was a crucial phase in formulating policy and recommendations that would eventually become law. Those recommendations were based on testimony from a host of stakeholders, yet it seems that only those made by representatives of the cultural industries had an impact. For example, David Rothman, a freelance writer, responding to the initial call for comments on the Green Paper, noted: "The user must have unrestricted access to information available through the National Information Infrastructure (NII). Users must not be inhibited in any way from any use of available intellectual property. The primary user concern here is for the FREE SPEECH RIGHT of use. There shouldn't be any restrictions on use; nor should there be any administrative burden placed on the user to limit uses to what seems 'fair'" (in *Comments* 1993, emphasis in the original).

Importantly, Rothman also critiqued the nature of the hearing and its composition, noting that the opinions presented by participants may not have captured a broader view of what the NII would mean for average users. Writing to the working group after the hearings, he said:

> I recall your Nov. 18th hearings as valuable and very well organized, but not as reflecting the needs of society at large. Industry witnesses clearly set the tone. . . . Even some school affiliated people were not always representing the true public interest. Certain educators for example did not seem to care that much about the cost of knowledge as long as teachers and students could dial up books through special licensing arrangements. Such people ignore the fact that most learning today takes place long after graduation. (in *Comments* 1993)

Rothman is generally correct. Comments from private citizens who imagined a different kind of impact by the NII (one that affected private learning, creative usage, or a need for an open system that would buttress free speech) were few and ultimately ignored in the Green Paper's recommendations. Images that struck a chord with policymakers saw industry as the purveyor of value and, ironically, of free speech. Steven Metalitz painted a picture of who would provide useful content: "If copyright cannot be protected in the new information environment, then the supply of useful information will be drastically curtailed—or just as troubling, it will be limited to information that government or some other powerful institution chooses to create" (in *Comments* 1993)—an ironic statement because media corporations are themselves powerful institutions deserving of the same healthy skepticism that government receives and also because in the digital movement it is these same industries that have been shown to be enemies of free expression as a result of stringent protection of cultural goods.

Rothman was also correct in noting the role of institutional actors in representing a more balanced approach. The presence of institutional actors such as libraries and universities as the sole representatives of rights to access colored how access was discussed, imagined, and given value. The discussions focused on preserving these institutions as conduits of knowledge, but no thought was given to those who might work outside of such structures. This omission yielded a significant blind spot in the discussions. Representative users who might choose to be more participatory in media consumption were not acknowledged. The fact is that although content contribution by the masses was not yet a recognized phenomenon in mass media, it was nevertheless already taking place across the early digital networks of the 1990s. Usenet, iRC, and early Internet bulletin boards had a significant amount of content that was contributed and used by average users. The issues that the WGIP addressed were only those it saw as affecting copyright owners. The issues framed by early adopters and current users of the NII were not understood or not heard.

The next chapter discusses the reactions to the Green Paper from stakeholders and an increasing number of average citizens, intellectuals, and activists, some of whom are here understood as early intellectual contributors to the digital rights movement. The White Paper, a second draft of the policy proposal for the NII, was a further attempt to incorporate stakeholders, but as has been shown in other works and is evident from a review of the Congressional Record, little was done to incorporate expansive views of user participation (see Litman 2001).

3 Origins of the Digital Rights Movement: The White Paper and the Digital Millennium Copyright Act

A comparison of the initial recommendations in the WGIP's Green Paper to the final recommendations in what came to be called the "White Paper" and ultimately the DMCA suggests some important outcomes from the policymaking processes for the DMCA. First, it shows the emergence of a number of visions of what the NII would become. During the comment period prior to the release of the White Paper, a number of citizens voiced concern over the possible excess of the proposed policy and suggested that the NII might be a place where copyright and intellectual property can be reimagined rather than reenforced. Second, such a review, specifically of the White Paper's final policy recommendations and the DMCA's most resisted provisions (the anticircumvention provisions), finds that, by and large, citizen concerns and imaginings were ignored. This lack of notice prompted initial resistance to the law and served as a sort of spark to the digital rights movement, giving early leaders a foil against which they might rally supporters. Last and perhaps not surprisingly, the comparison suggests that the policymaking process's approach to evaluating copyright policy favored the status quo argued for in the rhetoric of preserving the "balance of copyright." I ultimately ask whether technological changes alongside changes in consumer/user practices should not be the impetus to reevaluate the status quo, changing it to facilitate participation rather than to preserve or buttress restrictive copyrights.

Testimony and Hearings for the White Paper

Following the release of the Green Paper in 1994 (see WGIP 1994), the WGIP requested comments from stakeholders regarding its proposals. It received more than 140 written comments from copyright owners, trade associations, libraries, and individuals (see *Comments* 1994 and appendix B). About 47 were from individuals or organizations representing copyright interests, 27 were from libraries and universities, 27 from private citizens, one from a consumer advocacy group, and the remainder from Internet service providers (ISPs) and foreign associations.

The major difference between the comments before and after the release of the Green Paper was the representation of average citizens, mostly members of computer professions, online communities, and law professors. Their contributions to the policy process put forth alternative conceptions of the NII, the consumer, and what ought to be the limits of digital copyright. Not surprisingly, copyright owners overwhelmingly agreed with the Green Paper's conclusions and suggestions: that only "minor" changes were needed in the copyright statute to ensure continued strong copyright protection in the NII. They also noted that technological measures in conjunction with the law would help ensure this protection and that technology that circumvented technological enforcement ought to be outlawed (see *Comments* 1994 and *Testimony* 1994).

A number of business and public-interest groups raised strong objections to the Green Paper's recommendations, however. ISPs, for example, argued that holding them strictly liable for their users' potentially infringing activities would stifle their nascent businesses. They claimed that the anticircumvention provisions and the strict liability for ISPs proposed by the Green Paper would hamper innovation, market expansion, and free speech. They noted that forcing ISPs to monitor users would impose a crippling economic burden, making it difficult for the industry to grow. It would also create an element of surveillance in the users' experience of the NII and negatively impact free speech.[1]

The strongest critiques of the Green Paper's recommendations came from individuals and law professors in support of a more user-centered view of copyright. Of the more than 140 written comments submitted to the WGIP after the release of the Green Paper, 20 were from private individuals concerned with the impact of policy on personal use. These individuals noted that technological enforcement and notions of fair use governed by licensing were counter to user experience with digital media, would have potentially negative impacts on innovation, would defeat the purpose of the NII's "network" effects (which gives value to information because it is distributed), and would be unenforceable in the NII's distributed networks.[2]

Many of the individuals who commented after the release of the Green Paper were computer professionals, such as computer science professors and programmers, and they deployed arguments common among hackers, who view proprietary claims on software as ultimately detrimental to the common good. Also, many respondents framed their anti–Green Paper comments in notions of a "proper balance," noting that the Green Paper had missed the intent of the intellectual-property clause in the Constitution. One such witness noted that

> the draft report appears to have concerned itself mostly with the issue of protecting the property rights traditionally granted to copyright holders, and how to stem the tide of change that is rendering them unenforceable and archaic. Not really addressed is how best to promote the "progress of science and the useful arts," the real reason for copyright and patent protection in the first place. For example, the software that runs NII as it exists now (i.e. the internet [*sic*]) was constructed without substantial protections for the ideas that underlie it. I think that if the original authors of the ftp program had patented the ideas that went into it, it is more than likely that I would have been able to use it to get the NII-ip document from your computer. Thus, the interest of society was best served by their not protecting their property and their not interfering in other people's use of it. (Mitchell Golden in *Comments* 1994)

Even stronger critique came from law professors who could speak to the interpretive biases in the Green Paper's recommendations. Among them, none was as scathing as Jessica Litman of Wayne State University Law School. Noting that the Green Paper disturbed the balance between the public good and incentives to authors, she took particular aim at copyright owners' claims that fair use was simply a defense against liability of infringement and implied that fair use has in fact been viewed as a right. She added: "The public does not give out copyrights to encourage authors to appropriate all of the rents that a given creation might yield. The copyright system is designed instead to assist authors in earning enough profits to enhance the creative environment enough to make their works available to us" (in *Comments* 1994).

In her view, the primary purpose of copyright is the good of the public, not the good of the author. This view strikes at the heart of the differences between notions of fair use and copyright held by copyright owners and proponents of expanded fair use. Furthermore, Litman conceptualized a different sort of consumer than had been envisioned by either copyright holders or the supposed representatives of the public—educational institutions and libraries: "The library associations are here purporting to speak for the public. And they surely speak for the public's ability to use the NII in libraries," she wrote. "But nobody involved in drafting this report appears to have seriously estimated the interests of the public in general" (in *Comments* 1994).

What are the interests of the general public? Fostering creativity, Litman argued. She, along with other proponents of expanded fair use, believed that allowing users access to copyrighted works will "benefit all of us in a variety of creativity enhancing ways" (in *Comments* 1994).[3] She disagreed with recommendations that would limit first sale in digital media or that would regulate every use of digital products. To Litman and others, the consumer was much more than simply a consumer; he or she was a potentially creative appropriator who would need expanded access to copyrighted works to use them in ways that would allow for creating new works. This definition of the user is a central vision for what other legal scholars have termed the "situated user," one that is not simply limited to consumption but has creative rights as well (J. Cohen 2005).

The notion of the "situated user" has been one of the rallying cries for the digital rights movement, and it was a failure on the part of the WGIP not to recognize users as more than consumers. One user explicitly noted,

> Much of the value on the internet [*sic*] has come from the blurring of lines between providers and consumers. The content which has built the Internet so far has come almost entirely from peer-to-peer interactions. . . . A substantial amount of high quality content has emerged . . . including libraries and archives, award winning periodicals and top-quality soft-ware. . . . The tendency among some industry analysts has been to dismiss such benefits as "minor league action" in comparison with traditional commercial systems. But to some extent the relatively youthful Internet is already challenging that presumption. Why is it growing at such an explosive pace if, as the WGIP suggests, the content to drive its success is not yet in place? (Mahatma Kane-Jeeves in *Comments* 1994)

Based on such testimony, the WGIP should have been aware that at least some of the public had a different perspective on the origins and value of content on the NII—namely, that it was user created. Yet in the WGIP's final recommendation, it disregarded the notion of active and creative users whose practices were already entrenched. This choice resulted in an adversarial relationship between the law and many users of digital content because the WGIP recommended not extending first sale to the NII, defined fair use narrowly, and advocated for criminalizing circumvention technologies.

Alternate Visions of the NII

The critiques of the Green Paper's recommendations by consumer rights groups, law professors, and private individuals articulate an alternative vision of the NII. The rationale for strong copyright protection on the NII, articulated by the WGIP and content owners, assumed that the NII would be barren without big commercial contributors and saw the challenges inherent within the NII as threats. Others, however, believed that there would be value nonetheless or that for value to be truly exploited the NII should not be a place with strong copyright protection, but instead a place where there may be weaker protection that would allow users maximum access to the potential of digital media and networks.

The latter visions went beyond legal arguments, and for users of the NII they were rooted in technological practices that content owners and the WGIP did not understand. Many people who commented on the Green Paper could not see the rationale espoused by content owners for why there should be strong copyright protection on digital content. As one software developer put it, "When duplication is not difficult, many people rightly ignore the shady government granted monopoly that we call intellectual property. After all, only fools think [intellectual property] is tangible and no amount of prattle from lawyers will really convince anyone that copying a computer file is the same as taking someone's diamond necklace. Who am I to say this? Just a private citizen who happens to write software for a living" (Richard Johnson in *Comments* 1994).

Another witness wrote: "The moral implications of copyright violation on the nets are far from clear. Simplistic analogies to stealing and inflated claims of software companies notwithstanding, the true dimensions of harm and losses from common small scale piracy have never been well established. It is not unreasonable to speculate that many instances of illicit copying do not in fact result in any actual harm to the copyright owner" (Mahatma Kane-Jeeves, in *Comments* 1994).

These comments suggest that copying of intellectual property was seen as a common practice, and although I do not suggest that authors' rights ought to be ignored, the issue should have been addressed with more finesse than an all-encompassing focus on enforcement through technological protection mechanisms and criminalization.

Both camps continued to view technological enforcement of copyright differently after the release of the Green Paper. Most notably, individuals experienced with computers warned the WGIP that technological enforcement would be difficult. They argued that encryption schemes would only create a speed bump against unwanted access and copying and that enforcement mechanisms would continuously be defeated. Richard Johnson's comments suggest that even potential copyright owners were in the habit of copying and had the skill to write software to help them do so. David Rothman, quoted in chapter 2, foresaw the kinds of digital civil disobedience and resistance to law and technology that one day might come about if the WGIP's recommendations were accepted: "If the

Lehman abomination becomes law, we just might see National Copywrong Day—during which Netizens could mail each other copyrighted articles and publicly announce their sins to mock Washington" (in *Comments* 1994).[4]

Without exception, private individuals commenting on the NII wanted to see a network where information was easily accessible and not encumbered by law or access-control technologies. And whereas libraries and many lawyers saw the WGIP's recommendations as legally problematic, users saw them as counter to the common practices of life in the new medium, the NII.

The WGIP made no concessions to these views in the White Paper, the final draft of its proposed recommendations, noting:

> While, at first blush, it may appear to be in the public interest to reduce the protection granted works and to allow unfettered use by the public, such an analysis is incomplete. Protection of works of authorship provides the stimulus for creativity, thus leading to the availability of works of literature, culture, art and entertainment that the public desires and that form the backbone of our economy and political discourse. If these works are not protected, then the marketplace will not support their creation and dissemination, and the public will not receive the benefit of their existence or be able to have unrestricted use of the ideas and information they convey. (WGIP 1995, 14)

The paper also stated, "Since computer networks now make unauthorized reproduction, adaptation, distribution and other uses of protected works so incredibly easy, it is argued, the law should legitimize those uses or face widespread flouting. . . . Computer networks can be and have been used to embezzle large sums of money and to commit other crimes. Yet, these acts are prohibited by law. Simply because a thing is possible does not mean that it should be condoned" (WGIP 1995, 15).

There are a number of difficulties with these responses. The first response is rooted in the rhetoric of incentives, which assumes that all incentives to produce creatively are monetary. It also assumes that a marketplace of goods based on strict copyright privileges is the only marketplace capable of generating value and profit. However, this is not always the case. The realization of the open-source/free-software movement as a profitable enterprise has shown itself to be an exception to this assumption, and other examples abound. Building primarily on freely available products, open-source ventures have generated revenues that come from services provided in support of the product's use. Production of open-source software was originally not motivated by monetary compensation, but rather by norms of gift giving and reputation on the Internet.

Also, alternative licensing systems such as the Creative Commons license (discussed more fully in chapter 8) have shown that authors are willing to give users more rights to access and use and that giving such rights does not preclude their incentive to produce. In short, the WGIP turned a blind eye toward the potential for "participatory culture" among users (see Jenkins 1992; Lasica 2005), incentives to authors, and the nature of alternative sources of revenue on the NII. It was influenced by the alarmist arguments espoused by content industry representatives describing the death of the industry on the NII and indeed overly anxious to craft policy on their terms. As the proceedings show, industry representatives used well-worn tropes to craft their arguments about the appropriateness of strong copyright protection. Painting a picture of sublime creators toiling to realize the genius of their ideas, the industry argued for their protection against those who would steal their intellectual property and for incentives to spur their creativity forward (see Boyle 1996; Patterson 1968). Even a cursory look at today's culture industry reveals that the majority of owners of copyrighted works are media businesses; that the lonely genius is but one player in an interconnected system of production and consumption that includes the cultural commons, business, creators, and consumers; and that the latter two are increasingly converging.

The second response quoted earlier is problematic because it oversimplifies the nature of copyright by ignoring the categorical difference between the nature of copyright's social contract and tangible property as protected by law. They are not the same. Copyright is a limited property right: it has limitations with regard to the fair-use doctrine; it has limitations with regard to the exemptions for libraries and the interoperability of technological systems; and it has limitations on control of distribution, such as the first-sale doctrine. Thus, property rights in intellectual property are supposed to be in a state of negotiation where the interests of the public are the ends of the doctrine and monopoly for authors is the means. If a technology encroaches on copyright owners' rights, the questions regarding policy *should not* be limited to concerns over maintaining or reestablishing the balance of copyright but should also ask *whether the new copyright relationships made possible by the emerging technologies will actually be more beneficial to the public*. In other words, given changing technological regimes, we should not shy away from considering whether copyrights should be changed in favor of user access. The trend has always been for preservation of copyright, but who is to say that a decrease in the rights granted to authors would necessarily be detrimental to society? Rights of ownership in intellectual property are, more than any other right, clearly granted by the state through the beneficence of the public and as such can be subject to constant revision and reevaluation (see Boyle 1997).

In sum, comments and testimony after the release of the Green Paper were largely critical if originating from technology users and technology firms invested in the free flow of information on the NII but were supportive if coming from content owners. As will become apparent, the vision of the NII that earned a place in the final draft of the WGIP's report on intellectual-property issues on the NII was that espoused by content owners.

The White Paper: A Second Attempt at Formulating Policy for the NII

The White Paper was released in September 1995 and has been widely criticized as revisionist in its interpretations of copyright law. Law professor Jessica Litman (2001) has noted that Commissioner Bruce Lehman responded to negative comments primarily by ignoring them or by noting that "naysayers" simply did not know what they were talking about. Furthermore, critics suggest that the White Paper, in response to negative comments on the Green Paper, focused on reinterpreting case law and legislative intent to suit its purposes. As Litman has noted, the majority of the White Paper was geared toward grounding its recommendations in legal doctrine to make them appear less revolutionary in their impact on the scope of copyright. Overall, the White Paper's recommendations and its rationale for them were the same as those in the Green Paper (Litman 2001).

The White Paper differed with the Green Paper in two important respects concerning its recommendations for technological protection measures and law, however. First, it supported changes in the standards used for criminal liability and legislation that would become the No Electronic Theft (NET) Act of 1997.[5] Second, it changed its recommendations for amendments to sections 501, 503, and 512 of the US Copyright Act, as noted earlier, recommending instead the addition of a new chapter 12 to US Title 17. Chapter 12 would serve more or less the same function as the previously proposed amendments.

Technological Protection Measures and the Law

The White Paper considered the types of technological protection measures that could be available to content owners who chose to distribute digitally. The measures included: (1) access controls at the level of servers and files, essentially password protection for access to a file or a server holding content; (2) encryption of data with distribution of decryption keys to paying consumers; (3) digital signatures or embedded code, usually in a specific location in the digital media that validates the authenticity of the work and ensures that if the work is copied without authorization, subsequent copies can be identified; and (4) steganography or watermarking technology embedded code in digital media that will prevent them from being rendered if the watermark is removed or tampered with and that can also serve as an authentication measure similar to a digital signature. In its discussion of these measures, the White Paper noted that any combination can be used to ensure that copyright is maintained in digital works.

When the White Paper considered unlawful distribution of proprietary content online, it stated that legal remedies should be strengthened to criminally sanction those who willfully infringe on copyright even *if those infringing or contributing to infringement do not gain financially from it.* The WGIP cited the 1994 court case United States v. LaMacchia (871 F.Supp. 535 [1994]), where a lawsuit was brought against David LaMacchia, an MIT student and hacker, for posting copyrighted material on an electronic bulletin board. In that case, the court was forced to dismiss the government's attempt to bring criminal penalties against LaMacchia because the law did not provide for criminal liability for individuals distributing copyrighted works for no financial gain. The White Paper recognized that "[s]ince there is virtually no cost to the infringer, certain individuals are willing to make such copies (or assist others in making them) for reasons other than monetary reward. For example, someone who believes that all works should be free in Cyberspace can easily make and distribute thousands of copies of a protected work and may have no desire for commercial advantage or private financial gain" (WGIP 1995, 229). Without noting it and perhaps without knowing it, the White Paper was describing the more radical elements of the open-source/free-software movement as well as intimating that some other form of reward mechanism may exist among Internet users—a form of reward that encourages distribution of copyrighted works.

The solution, from the White Paper's perspective, was a natural knee-jerk response to a complex behavior rooted in norms and user practices: to increase criminal liability. Perhaps the better response would have been to ask why otherwise perfectly law-abiding citizens might come to believe it is acceptable to violate copyright on the Web. This approach would have required a conceptual shift in the way content owners and the WGIP thought of copyright (admittedly a shift against industries' interests, but perhaps toward the interests of society as a whole). Let me not be misunderstood. To ask the question "Why might otherwise perfectly law-abiding citizens come to believe it is acceptable to violate copyright on the Web?" is not to entertain the notion that just because users can do such a thing, it should now be considered a legitimate option for policy and law. Rather, to pose the question suggests an understanding that the notion of intellectual property is a bargain between society writ large and individuals, a bargain whose terms are dependent on technological, historical, and cultural circumstances. If the circumstances begin to shift, it is understandable to be conservative and preserve the terms (in this case the law), but it is perhaps more important to be open to changing those terms to include new groups of people and their practices. It is a tricky balancing act to include in the discourse those who might easily be labeled criminals for their behavior, but if policy is to grow with society's needs, the policy process must include them. The significance of the White Paper's response to LaMacchia is that it demonstrates a particularly conservative (and perhaps overly simplistic) way of addressing a complex issue of law and social behavior.[6] From a technology perspective, the White Paper sought to curb behavior such as LaMacchia's by buttressing protection measures with the power of law.

The White Paper noted that legislation should ban the manufacture and use of circumvention technologies. Such an action was not unprecedented in technology policy; as noted earlier, both the Telecommunications Act and the AHRA outlawed the circumvention of technological measures controlling copying of transmitted signals and audio recordings.

Much like the Green Paper, the White Paper continued to ignore concerns that technological protection measures would limit fair use, and anticircumvention provisions in law would give them legitimacy online. The White Paper reiterated its position that fair use is not a right that content owners are required to allow for. As such, in this view, technological protection measures and anticircumvention law are well within the scope of the copyright statute. As noted earlier, consumer practices and notions of fair use were not considered in the WGIP analysis of fair use, nor was the power of users and hackers effectively to undermine the White Paper's technolegal regime with hacks or work-arounds.

The DMCA

Following the comments on the Green Paper, the WGIP released the White Paper (WGIP 1995), the final draft of its recommendations for copyright on the NII. Then, based on the recommendations and continued consultation with stakeholders, Congress formulated and passed the DMCA in 1998.

The DMCA is an amendment to US Title 17, the US copyright law, bringing it in line with the World Intellectual Property Organization (WIPO) Copyright Treaty (WCT) of 1996 and the WIPO Performance and Phonograms Treaty (WPPT) of 1996. Both of these treaties were expansions of the Berne Convention for the Protection of Literary and Artistic Works of 1886 undertaken in recognition of the new threat to copyright that emerging digital technology posed.[7]

Specific sections of the DMCA have come to be known as its "anticircumvention provisions." They were informed by the White Paper, implemented with the support of US copyright owners and under requirements from the treaties mentioned in the previous paragraph. As the US Copyright Office noted,

> Each of the WIPO treaties contains virtually identical language obligating member states to prevent circumvention of technological measures used to protect copyrighted works, and to prevent tampering with the integrity of copyright management information. These obligations serve as technological adjuncts to the exclusive rights granted by copyright law. They provide legal protection that the international copyright community deemed critical to the safe and efficient exploitation of works on digital networks. (1998, 3)[8]

The anticircumvention provisions comprise DMCA Title I, section 1201(a) and (b), and they are the legal embodiment of conceptions of how access to digital content should be structured. They place the consumer under the tight control of technological measures and criminalize the design and use of technology that might give consumers extended or unauthorized use. Many of the court battles that have been fought over violations of copyright since the late 1990s have revolved around circumvention of the technologies protected by section 1201. The digital rights movement has focused in part on challenging the technolegal regimes that these sections create. For these reasons, I spend a bit of time reviewing section 1201.

This section of the DMCA was based on the premise that copyright owners should be encouraged to help themselves by creating technological measures that would ensure their copyright is preserved. This notion was fostered by a generalized rhetoric of "fear and consequences" for copyright on the NII. Thus, the WGIP and subsequently Congress intended to give copyright owners the broadest protection against consumers, who were envisioned as passive receivers of information from the NII or, worse, as potential thieves. To achieve this end, section 1201(a)(1)(A), first, prohibits the conduct of circumventing technologies that control access to copyrighted content, thus making "cracking" or breaking a technological lock illegal. Second, the statute prohibits the manufacture and distribution of technologies that might help in carrying out circumvention. Section 1201(a)(2)(A), (B), and (C) state:

> No person shall manufacture, import, offer to the public, provide, or otherwise traffic in any technology, product, service, device, component, or part thereof, that—
>
> (A) is primarily designed or produced for the purpose of circumventing a technological measure that effectively controls access to a work protected under this title;
> (B) has only limited commercially significant purpose or use other than to circumvent a technological measure that effectively controls access to a work protected under this title; or
> (C) is marketed by that person or another acting in concert with that person with that person's knowledge for use in circumventing a technological measure that effectively controls access to a work protected under this title.

Thus, section 1201 outlaws hacking into protected media as well as developing the tools associated with hacking. Furthermore, it has a great stake in the deployment of anticircumvention technologies because it implicitly accepts the impracticality of enforcing copyright law in the digital domain with traditional law enforcement methods.

Section 1201 separates its understanding of technological copyright-protection methods into two categories: (1) measures that control access to a work and (2) measures that control the exercise of exclusive rights with respect to a work. The first type of protection is called "access control," and the second type is often referred to as "copy control." The reason why legislators sought to create this distinction between protection methods is that fair use gives consumers certain privileges over the works that they have purchased (see the section on fair use in chapter 2). Technologies that control the exercise of exclusive rights (copy-control technologies), legislators reasoned, also bear on consumers' fair-use privileges, and it would be inconsistent to hold consumers liable for circumventing technologies that preclude them from exercising those privileges. Thus, section 1201(b) purposefully does not state that the conduct of circumventing copy control technologies is illegal. However, section 1201(b)(1)(A), (B), and (C) state that the manufacture and distribution of these technologies *is illegal*. Thus, whereas one part of 1201(b) implies that the conduct of circumventing copy-control technologies is allowed because these technologies control privileges that the consumer may have over purchased works, another part says that any technology that may help the consumers exercise those privileges cannot be distributed.[9] The consumer, if she is savvy enough, can certainly design a piece of software to circumvent copy-control technologies. However, if she is not, then she cannot exercise fair use by circumventing copy control because the prohibition on distribution prevents her from being able to buy the technology, get it from a friend, or download it from the Internet. Section 1201(b) makes allowances for the conduct of circumvention hollow because the majority of consumers do not have the technical know-how to design copy-control circumvention technologies on their own.[10] This provision is thus a weak technological solution for the implications of digital protection technologies with regard to fair use. By feigning an allowance for circumvention technology of copy control, the WGIP satisfied itself that it had kept fair-use interests in mind.

The distinction between access-control technology and copy-control technology is important for copyright owners. Access to a work is the sole privilege of the owner of the work and is not protected by fair use or any other exceptions to the author's exclusive rights. Therefore, any copyright-protection technology that controls access to a work is essentially like a locked door at a private residence: one may not enter unless invited. The provisions in section 1201 regulating access and the technologies that ensure it illustrate how technology and the copyright statute have come together to redefine the relationship between copyright owners and consumers. Access to a work is a negotiable condition of purchase. When consumers buy works protected by access technologies, they are not really buying the works in a sense that would grant them fair-use privileges. Rather, they are buying access, and the terms of sale may include not only a rent, but the surrender of fair-use privileges and first sale. Such surrender is difficult to enforce and implement with nondigital materials; in other words, the possibility of it being a term of sale/rent is afforded in relation to digital technologies and materials.

Take the following example. A consumer can buy a book at a bookstore. Implicit in that exchange is the purchase of both access to the work and the assumption by the consumer of certain fair uses. Publishers would much rather lease the consumer the book and charge a rent every time the book is used, but market and technological conditions make such a condition of sale impossible. Publishers must sell the book, accepting the technological and market limitations that prevent them from leasing it out to the customer. The copyright owner must make the good-faith assumption that the consumer will not overstep her fair-use privileges primarily because the publisher has no practical way of controlling the personal uses a consumer makes of a printed book. It is imaginable that the copyright owner of the book, in fear of his exclusive rights being abused, can try to negotiate the sale of the work in exchange for the consumer's surrender of fair use, but how can the surrender of rights be enforced? It would be impractical and absurd, for example, for booksellers to follow consumers everywhere they go, making sure that they have not violated the contract. There is nothing one can do to the printed book to ensure that it regulates the contract for itself. Thus, given the limitations, access to a work in print is unencumbered with regulatory mechanisms. The law delineates the rights of authors and the consumer privileges, but technological limitations and market realities (Who would purchase a book with an armed guard attached?) keep the boundaries of the agreement between buyer and seller from sliding into the realm of the absurd. The conditions of the bargain are more or less a practical matter.

Digital technologies have made overcoming the impractical trivial, however. When one purchases an eBook, downloads a song from one of the many online music stores, or purchases a DVD, it is possible for access technologies to come with the package. It is also possible for access to be negotiated in a EULA and for the license to limit consumer uses, effectively requesting that the consumer rent a work or surrender fair use. It is in copyright owners' interests to distribute only works protected by access-control technologies because these technologies give owners the broadest possible control over their media.

Some would argue that this vision cannot be realized because consumers can go elsewhere for their content if they find access-control technologies and the associated agreements too restrictive. However, the claims ignore the market power held by copyright owners. Because owners have exclusive rights to license distribution of their content, all vendors can be required to have access-control technologies in place. In that case, the consumer may truly have no place to go for a better deal.

In recent years, the major sellers of online digital content have made some changes—selling, for example, music that has no DRM system associated with it (Apple being the most notable, but also Amazon.com). Such a change is in no small part due to consumer demand and the discursive inconsistencies of marketing campaigns that sell not only music, but also the idea of a personalized media experience and then turn around and limit or disallow personal noninfringing uses. Despite this turn to more unencumbered digital media, the vast majority of digital content (books, movies, videogames, software, and streamed content) still remains under tight technological control.

Initial Resistance

By the time the DMCA was passed into law in 1998, a host of prominent legal scholars and Internet "gurus" had started to point out its biases and inconsistencies. As noted earlier, some critiques came during the formulation process from users of the Internet; other critiques came from libraries addressing their need for continuance of privileges held in the analog world into the digital world. Still others were formulated by ISPs and consumer electronics groups who wanted to ensure their business models, arguing that overprotection would stifle innovation. Many of these critiques were ignored.

An important development in the movement against expanded copyright protection on the NII was the involvement of a small circle of elite legal scholars and technologists who early on forcefully politicized the implications of digital copyright. They include John Perry Barlow, founder of the Electronic Frontier Foundation (EFF), and law professors James Boyle, Jessica Litman, and Pamela Samuelson. I discuss the importance of Lawrence Lessig in the digital rights movement in chapters 4 and 8, but the first four were the earliest critics of the White Paper and the DMCA.

John Perry Barlow's contribution can be said to be primarily ideological. In a widely distributed article, Barlow noted: "We are sailing into the future on a sinking ship. This vessel, the accumulated canon of copyright and patent law, was developed to convey forms and methods of expression entirely different from the vaporous cargo it is now being asked to carry. . . . Intellectual property law cannot be patched, retrofitted, or expanded to contain digitized expression" (1994).

Barlow's connections to the EFF positioned the organization to become a leader in the movement against expanded copyright. He also achieved notoriety among the cyberlibertarian communities of cyberspace, authoring the infamous "A Declaration of the Independence of Cyberspace" (Barlow 1996). He espoused minimal government intervention within cyberspace, imagining a cyberutopia of free information flow and total equality. Critics, of course, have of course pointed out the shortfalls and realities that hinder this vision, but it captured the imagination of many early Internet users. It questioned the "unquestionable" claims of ownership of cultural products espoused by publishers and copyright holders and challenged governments' authority to regulate the communities taking shape in cyberspace. In many ways, his work and the work of others (such as Richard Stallman [2002]) politicized the Internet and made it a contested space where users and content owners (often one in the same) as well as governments were forced to reconsider how and who was actually being regulated by policies such as the DMCA and the expansion of copyright to the digital world.

Barlow's critique of intellectual property in digital works was expansive; he questioned the authority of an international intellectual-property rights regime fashioned by Western nations, the feasibility of regulating copyright on international data networks, the effects on free speech and innovation that such regulation might have, and the divergence between common practices in the digital domain and what the law (along with technological enforcement) was now capable of doing. Barlow also wondered about the nature of ownership in digital products that can be reproduced and distributed at nearly null cost and leave the original owner no less the richer from the distribution. Although he did not advocate that copies of original works ought to be given away for free, he wondered whether current prices were fair and presciently predicted the ongoing legal war over digital copyright. Barlow's contribution was visionary and at least in part predictive of alternative methods of extracting value from intellectual property.

> Perhaps those who are part of the problem will simply quarantine themselves in court, while those who are part of the solution will create a new society based, at first, on piracy and freebooting. It may well be that when the current system of intellectual property law has collapsed, as seems inevitable, no new legal structure will arise in its place.
>
> But something will happen. After all, people do business. When a currency becomes meaningless, business is done in barter. When societies develop outside the law, they develop their own unwritten codes, practices, and ethical systems. While technology may undo law, technology offers methods for restoring creative rights. (Barlow 1994)

The open-source movement and the success of open-source software as a business model bear out his predictions. Beyond that, his thoughts on the nature of the Internet, specifically in terms of its situated norms and its possibility to maximize the value of information by making it widely available, informed critics of digital copyright who saw legal strictures as impediments to accessing information and cultural products that are at least in part components of a cultural commons.

James Boyle was one of the first legal scholars to call for a movement to reappropriate cultural products that were quickly being put out of reach by the machinations of copyright owners. His generative 1997 article called for a social movement and a discourse to save the quickly diminishing commons: "Right now, we have no politics of intellectual property—in the way that we have a politics of the environment or of tax reform. We lack a conceptual map of issues, a rough working model of costs and benefits and a functioning coalition-politics of groups unified by common interest perceived in apparently diverse situations."

His argument was rooted in the economy of information, which necessarily is defined by a balance between costs of information hording and the cost of information sharing. From Boyle's point of view, the cost of hoarding information and of intellectual-property protection outweighs the cost of information sharing. The prevalent rhetoric that equates benefits to authors with benefits for society informs the framework that has led policymakers to overprotect content rather than consider the benefits of underprotection. Boyle also was concerned with the privatization of speech by copyright, which circumvents free-speech protections and leads to corporate censorship.

When the White Paper was released, Boyle targeted it specifically as the embodiment of a policy position that would greatly help copyright owners circumscribe the commonwealth of culture and faulted it for not considering the implications of digitization for ease of use and efficiency, which might warrant "underprotection" to maximize the NII's network effects. His arguments attacked the prevalent perception of intellectual-property rights as "natural rights," noting that these rights above all others are not natural but granted temporarily by the state and as such ought not to reside within the protection of dominant rhetoric that holds other property rights as sacred and inviolate. In his early writings, Boyle touched upon the major issues now considered to be germane in the digital rights movement: fair use, creativity, innovation, privacy, and free speech.

In March 2006, Copyfight and Stanford's Center for Internet and Society acknowledged Boyle's influence on informing movement advocacy, holding a conference to gauge the progress of what they termed the "cultural environmentalism" movement. Ten years after Boyle's famous article, his views were adopted by Lawrence Lessig (until recently one of the digital rights movement's most visible intellectual leaders), whose Creative Commons license acknowledges the value of the cultural commons for continued cultural production and expansion while at the same time explicitly recognizing policymakers' failure to come to terms with the benefits of underprotection of intellectual property.

Pamela Samuelson was also an early critic of the DMCA, targeting its anticircumvention provisions as overly protective and criticizing the WGIP for its lack of interest in the unintended consequences of technological enforcement. Her key insight, one that I have expanded on here, is that the DMCA implicitly gives users the ability to circumvent copy controls but not the ability to get the software that would do the job. She asks "[w]hether it is lawful for people to develop or distribute technologies that will enable implementation of the exceptions and limitations on the circumvention ban built into the statute. Did Congress intend to allow people to exercise these privileges, or did it intend to render these privileges meaningless because the technologies to enable the excepted activities have been made illegal?" (1999, 46).

Samuelson's early work in writing about the inconsistencies of the anticircumvention provisions and her participation in conferences and symposia drew attention to the broad protections afforded copyrighted material in digital media and helped to propel this debate into the legal consciousness. Her early work has earned a place of prominence in the digital rights movement. The Boalt School of Law's Law, Technology, and Public Policy Clinic bears her name, and she sits on the boards of multiple key organizations and digital rights groups.

Perhaps no one has done as much as Jessica Litman to bring attention to the DCMA's excess and the historical trajectory of copyright in the twentieth century and into the twenty-first. Both by her testimony at the proceedings for the White Paper and with her writings, Litman was early to point out alternative visions of the NII and consumers therein. In her testimony before the WGIP, she criticized the committee for failing to represent the public interests and for catering to copyright owners. In her book based on her experiences with the formulation of the DMCA (Litman 2001), she notes that licensing schemes have been given dominion over fair use and personal noncommercial uses, creating a situation where consumers will potentially have to pay every time they want to access a work. She notes that the law and technology have allowed copyright owners to charge a rent for every potential use, a situation that is counter to the intent of the intellectual-property clause in the Constitution. Litman's work within the legal community, her participation on the boards of various key advocacy groups, and her publications helped galvanize others in the early days of the digital rights movement.

The White Paper and the DMCA created a technolegal regime that users of the NII and digital technologies immediately found restrictive. Policymakers marginalized many users by ignoring warnings concerning the loss of fair use, the common practices of digital technology users, and the problems with technological enforcement. By creating inconsistencies and biases in the law, policymakers created fertile ground for a mobilization against the law that seeks to regain control of media consumption, use, and access and to allow consumers to be more creative.

Part II discusses specific examples of the mobilization inspired by key prosecutions, lawsuits, and other types of repression that resulted from the DMCA. It uses case studies to illustrate how the movement shaped its important discursive frames, how some organizations rose to prominence, the movement's tactics, and the movement's organizational structure.

Part II

In this part, I leave the legislative history and analysis of the DMCA for an analysis of important cases in the development of the digital rights movement and a description of its dynamics and structure. The cases are meant to illustrate examples of repression and activism that have come to define the movement and to show how the movement actors coordinated with each other. A few important points emerge.

First, significant arguments about the movement's legitimacy can be seen in the cases presented. Fair use is a powerful theme for the movement, and it is deployed in a user-centered manner, tied to arguments of free speech. The strategy of framing the movement as part of broader "rights activism" is important in that it translates an issue that at the time of the movement's beginnings only marginally affected average media consumers into an issue that broader publics could potentially identify with.

Second, the cases and analysis of movement tactics and structure show the importance not only of organizations and intellectual leaders, but of hackers and activists who design technologies to facilitate access and use. The equal importance of these two groups is significant because it suggests that activism and mobilization need not rely solely on the resources and coordinating efforts of organizations and leaders: a hack to a technological protection mechanism can have a mass effect when distributed online. A lonely committed hacker who gets thousands of people to download his or her circumvention software can have the same or greater impact as a large organization with significant resources.

Last, this part illustrates the importance of technology as a site of activism (the Internet), as the means of activism (hacking), and as the focus of activism (technological protection measures). In all the cases described, technology plays this tripartite role. Activists often use the Web to organize themselves and spread information, use computer programs or hacks to work around and subvert technological protection measures on content, employ the Web to distribute those hacks, and target technologies and the laws that they find unjust.

4 Dmitry Sklyarov and the Advanced eBook Processor

The DMCA became law in 1998, and in the ensuing years a number of prosecutions mobilized activists to coalesce against it. This chapter discusses one such case, that of Dmitry Sklyarov and his crack of Adobe's eBook encryption. Although other cases that brought the issues of copyright to the forefront were more widely covered in the media (the Napster case, for example), the Sklyarov case is singled out here because it illustrates the emerging dynamics of the movement. Furthermore, it is a case that struck a particularly strong chord among activists. So although activists were certainly involved in advocacy in the three years immediately after the DMCA became law, the Sklyarov case was transformative—in other words, an event in the history of a social movement that "dramatically increases or decreases the level of mobilization" (Hess and Martin 2006, 250). It illustrates how vested audiences attempted to capture the meaning of Sklyarov's arrest, how Adobe tried to control the responses to the arrest, and how mobilization occurred in its wake.

Early Repression and Issue Definition

In January 2001, Adobe Systems Inc., authors of the ubiquitous ".pdf" (portable document format) Internet document format and the widely used publishing application Photoshop, released the Adobe Acrobat eBook Reader as a free download to Internet users. Adobe released its product in conjunction with the online book distributor BarnesandNoble.com, which made available a series of eBooks that were formatted to be accessible using Adobe's application. Adobe's eBook Reader allowed for increased graphical power and touted a "true to print" look for works rendered through its interface. Also included in the software was a built-in browser that would allow users to purchase content from distributors such as BarnesandNoble.com while within the application.

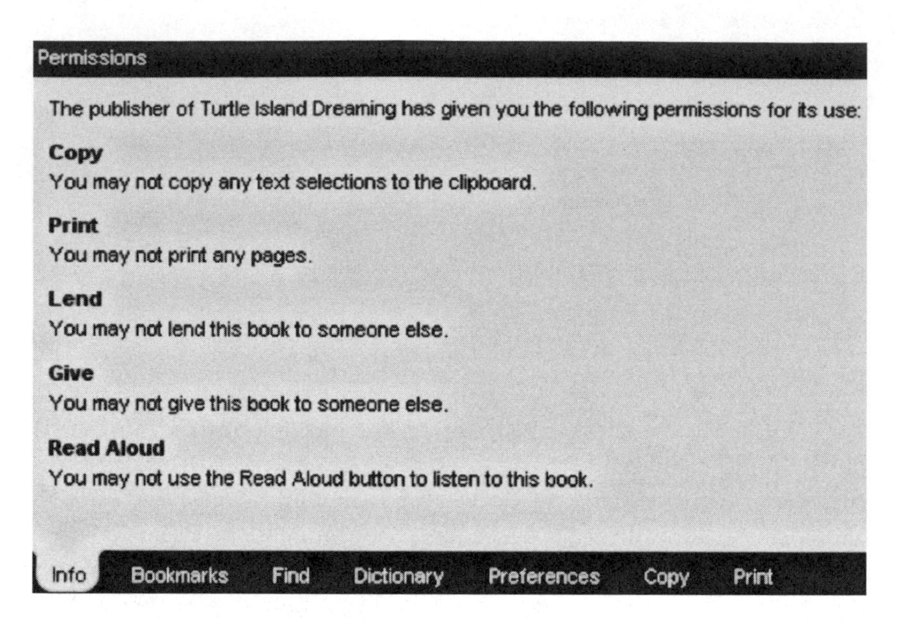

Figure 4.1
Permissions screen, Adobe eBook Reader.

The Adobe eBook Reader also came bundled with a DRM system. DRM systems for eBooks vary depending on the reader (such as Microsoft Reader, another popular reader). Regardless of the manufacturer, all eBook DRM systems work more or less in the same fashion: *the DRM system ties the eBook to the reader application that originally loaded it.* The DRM system's flexibility is determined by the publisher. For example, the Adobe eBook Reader allowed publishers to set permissions with regard to whether the books could be shared,[1] whether books could be printed, whether one could make copies or gift the book, and whether a user could use the Abode "Read Aloud Function" for the visually impaired (see figure 4.1).

Five months after the Adobe eBook Reader's release, the Russian software company ElcomSoft began selling the Advanced eBook Processor (AEBPR). The company specialized (and still does) in password-recovery software, and on June 22, 2001, it announced:

Advanced eBook Processor lets users make backup copies of eBooks that are protected with passwords, security plug-ins, various DRM (Digital Rights Management) schemes like EBX and WebBuy [included in Adobe eBook Reader], enabling them to be readable with any PDF viewer. . . . In addition, the program makes it easy to decrypt eBooks and load them onto Palm Pilot's [*sic*] and other small, portable devices. This gives users—especially users who read on airplanes or in hotels—a more convenient option than using larger notebooks with limited battery power to read their eBooks. . . . PDF protection can prevent users from changing or printing information, adding or changing annotations and form fields, or even selecting and copying text or graphics. With Advanced eBook Processor, these PDF files can be decrypted, opened, and used without any of these restrictions. Once protection has been removed, PDF files created with Adobe's Acrobat program can be opened in any PDF viewer, including Adobe's Acrobat Reader. (Katalov 2001)

Framed as a tool that could help legal owners of eBooks have more flexibility with the books they have purchased, ElcomSoft challenged the technological restrictions on the use of eBooks. ElcomSoft's AEBPR undid the DRM system on Adobe's eBook, which prompted BarnesandNoble.com to suspend sale of eBooks until Adobe made changes to the software so that AEBPR would not work. On June 25, just three days after the release of AEBPR, ElcomSoft received a cease-and-desist letter from Adobe. Adobe noted that the Russian company was in violation of the DMCA because it was distributing technology in the United States that allowed consumers to circumvent access-protection measures. Furthermore, Adobe contacted ElcomSoft's ISP, Verio Inc., and requested that the company's site be taken down. It also demanded that RegNow, ElcomSoft's fee-collecting agency in the United States, cease collecting fees for the sale of the AEBPR.

What followed was the beginning of a public-relations disaster for Adobe as the two companies exchanged public accusations on media outlets and hacker Web sites. When ElcomSoft finally went back online, Alexander Katalov of ElcomSoft accused Adobe of designing a weak protection system for eBooks and threatened to release the source code for AEBPR on the Internet under the protection of the free-software movement's General Public License (GNU GPL). In a post to ElcomSoft's new Web site and to the hacker newsgroup comp.text.pdf, Katalov angrily noted:

Now it's time for the brutal truth on Adobe eBook protection. We claim that ANY eBook protection, based on Acrobat PDF format (as Adobe eBook Reader is), is ABSOLUTELY insecure just due to the nature of this format and encryption system developed by Adobe. The general rule is: if one can open a particular PDF file or eBook on his computer (does not matter with what kind of permissions/restrictions), he can remove that protection by converting that file into "plain," unprotected PDF. Not very much experience needed. (quoted in United States v. Dmitry Sklyarov, US District Court [ND Calif., San Jose Div., 2001], Affidavit of Complaint, emphasis in original)

He also stated elsewhere, "[On whether Adobe's legal campaign will work] I should say that it will not work. We'll just move our site to another ISP, in another country (where there is no Digital Millennium Copyright Act [DMCA]). And/or make our software available for free, under the GNU license" (quoted in Planet eBook 2001a).

At the same time that Adobe was engaging in a war of words with ElcomSoft, it contacted the Federal Bureau of Investigation (FBI) office in Santa Clara County and informed the office of ElcomSoft's activities. It also told the FBI that Dmitry Sklyarov, who held the copyright on the AEBPR, would be in the United States for the hacker conference DefCon 9 in Las Vegas, Nevada, on July 15, 2001. In a sworn affidavit filed with the US District Court of Northern California, FBI special agent Dan O'Connell, following the information provided by Adobe, concluded that

Dmitry Sklyarov, employee of ElcomSoft and the individual listed on the ElcomSoft software products as the copyright holder of the program sold and produced by ElcomSoft, known as the Advanced eBook Processor, has willfully and for financial gain imported . . . a technology . . . primarily designed . . . for the purpose of circumventing a technological measure that effectively controls access to a work protected under Title 17, in violation of Title 17, United States Code, Section 1201(b)(1)(A) and Title 18, United States Code, Section 2. (United States v. Dmitry Sklyarov [2001], Affidavit of Complaint)

Thus, the stage was set for Sklyarov's arrest during DefCon 9, the most popular and well-attended hacker conference in the world. The FBI and Adobe could not have picked a more inopportune time and place for the arrest. Highly visible due to the press coverage leading up to his presentation at the conference and surrounded by tech-savvy friends and supporters, Sklyarov was arrested by FBI agents on July 16, 2001, as he was leaving the Alexis Hotel in Las Vegas. It took less than six hours for Internet media outlets to report his arrest.

The Transformative Nature of the Sklyarov Case

Sklyarov's arrest was immediately perceived as a repressive action on the part of the US government. In one account of Sklyarov's arrest, Katalov, an eyewitness, painted a picture of an innocent man being overwhelmed by agents of the state:

From July 11th to 16th together with colleague Dmitry Sklyarov, who was presenting a report, I attended the Defcon 9 conference in Las Vegas. On the morning of July 16th Dmitry and I left the hotel with the intention of going to the airport. We still had half an hour before the flight was supposed to leave when right at the front entrance to the hotel we were approached by two young men, yelling "Hands on the wall, FBI!" At first we thought this was somebody's idea of a bad joke (fed jokes were very popular at the conference). Dmitry laughed and tried to reply to the two men. The men, in a very rough manner, repeated, "Hands on the wall!!" A little bit later Dmitry was brought in wearing handcuffs. Dmitry asked to re-cuff his hands

> in front of his body as it was uncomfortable for him to sit down. The request was denied. The initiator of the judicial process was Adobe Software. The FBI men refused to give any further details saying that they were only following orders. . . . On my way to the airport I was trailed, very obviously actually. As soon as I tried to make a phone call in the airport a policeman ran up to a neighboring phone and pretended to call. He never did call anybody. (Katalov 2001)

This early account of the arrest was posted on the hacker Web site Slashdot.org. Framed as an overreaching act by state actors, the event quickly moved from a localized happening in relative isolation (a hotel lobby in Las Vegas) to a transformative event for the digital rights movement. The key to this transition was the construction of the event over the communication networks immediately available to those vested in the event. The fact that news of the arrest popped up in hacker sites almost immediately illustrates that this group found itself with a stake in the arrest. Hackers in particular took an active role in circulating information about it because one of their own was being prosecuted for an activity that all hackers undertake. They also had available to them a way to disseminate the information immediately (the Internet). Thus, hacker communities were the first to define the meaning of the event for broader audiences and used digital networks to beat most other outlets to the punch.

The first to break the story was the technology site PlaneteBook.com, followed the next day by Slashdot.org, the best-known hacker forum on the Internet. Within the next forty-eight hours, the news of Sklyarov's arrest would be reported not only by hundreds of Internet sites, but also by conventional media outlets such as the *New York Times*, *Pravda*, Reuters, ZDNet, MSNBC, CNN, CNET, *Wired*, and others. On Slashdot.org and throughout the hacker community, support for Sklyarov was strong. For example, on Slashdot.org almost four hundred comments were posted on the day following his arrest, of which almost all were supportive of Sklyarov. Many were critical of the DMCA's effects on free speech and technological innovation, and they were critical of Adobe's tactics. They called for reprisal against Adobe. Among the typical comments were:

> In its short life we have seen many security consultants and even college and university professors threatened with prosecution under DMCA for exposing weaknesses in computer security . . . activity which would otherwise be protected under the First Amendment and the traditions of academic freedom. (anonymous, quoted in "Fallout from Def Con" 2001)

> I have had entirely enough of this new adversarial stance of theirs. Let me just delete/opt/Acrobat4. . . . Their UNIX software sucks anyway. The rest of it isn't much better. Any software company that enforces or relies upon the DMCA should go on our blacklist! (anonymous, quoted in "Fallout from Def Con" 2001)

> I am writing to express my disappointment that Adobe would have a person arrested for pointing out flaws in one of its products. As a customer who cut my chops on Illustrator 1.1, it saddens me to think that Adobe now cares so little about the quality of its . . . products that it seeks to harass . . . those who point out their weaknesses. Some will call it "hacking" since it involved disabling a security routine, but I see it for what it is—pointing out a flaw in a product. Any company that would have someone arrested for protecting me can no longer enjoy my business. (D. Negro, quoted in "Fallout from Def Con" 2001)

The last of these quotes points to one of the most troubling consequences for Adobe resulting from Sklyarov's arrest. Due to the attention given to Adobe's eBook encryption as a result of the release of the AEBPR, security experts and hackers generally rejected it as a secure technology for copyright protection. The critique of the technology, in fact, was quite scathing. One hacker who had been quoted in the popular eBook technology site ebookweb.org, wrote, "How totally absurd PDF security really is. It is so weak and so lame that no self-respecting hacker or cracker would even bother breaking it. It simply isn't worthy of one's efforts" (Sperberg 2001). As Adobe's role in Sklyarov's arrest became apparent, the community of hackers commenting on the issue wondered whether Adobe was trying to cover up flaws in its eBook technology.

The speed at which support for Sklyarov emerged over the Internet cannot be overemphasized. By the end of the day on July 17, 2001, a little more than twenty-four hours after his arrest, the EFF, a central SMO working on movement issues, released statements on Sklyarov's behalf and began his legal defense campaign. Furthermore, the EFF directed members to the Web site freedmitry.org, where interested parties could contribute to a defense fund and get information on Sklyarov's condition by signing up for the "free Sklyarov" email list. Other Web sites protesting Adobe's role also appeared on the Internet. One such site, boycottadobe.com, organized a campaign to boycott Adobe products.

The EFF, along with community activists linked through freedmitry.org and its mailing list, articulated the frames through which the DMCA would be understood in Sklyarov's case. In a letter to Attorney General John Ashcroft dated July 20, 2001, the EFF noted that the case was one in which free speech, fair use, and innovation were being put at risk: "Now as law, it [the DMCA] is used as a powerful sword to squelch speech and competition and kill fair use. Congress never intended for the DMCA to destroy fair use, in fact it expressly tried to protect it. . . . [W]e ask that [as attorney general] you honor this intent and your obligation to uphold the Constitution by dropping the charges against Dmitry Sklyarov and allowing him to return home to his wife and two small children" (Steele 2001).

Activists and the EFF were also part of a group of movement actors who quickly organized collective action to protest the arrest. Just two days after his arrest, Sklyarov had a major advocacy organization working on his case and had served to mobilize collective action that would help define the digital rights movement as a movement with free speech and fair use as its central issues; and within less than a week a protest rally had been organized for July 23 in San Francisco.

The role of the media was also crucial in positioning the Sklyarov arrest as an important event in the movement. The media recirculated hackers' framing of Sklyarov's arrest as an indication of the DMCA's threat to free speech. Headlines read: "Case Highlights Law's Threat to Fair-Use Rights"; "Adobe Alerted Government to Russian Software Crack, FBI Agents Pounce on DefCon Hacker"; "Welcome to Vegas, You're under Arrest"; "Arrest Fuels Adobe Copyright Fight"; "Adobe Gets Russian Arrested"; "Another Russian Hacker Lured to US, Detained by FBI"; "Jailed under a Bad Law" (see PlanetPDF 2001 for the list of headlines).

Adobe, for its part, was unable to control the backlash against it stemming from Sklyarov's arrest. It could not refute or reframe the events leading up to and including the arrest because the events were so meticulously chronicled on the Internet and so quickly disseminated that it found itself trying to play catch-up as movement organizations and then news agencies portrayed Sklyarov as a poor unsuspecting graduate student ambushed while visiting Las Vegas and jailed as his wife and children waited back in Russia for his safe return (see figure 4.2).

Figure 4.2
Image of Dmitry Sklyarov and his wife and children distributed by the EFF, free dmitry.org, and freesklyarov.org.

Furthermore, the speed at which the network of activists coalesced was staggering and speaks to the Internet's coordinating ability and to the poor planning on Adobe and the government's part (see figures 4.3 and 4.4). As noted earlier, the government could not have picked a worse time and place to arrest Sklyarov. He was arrested shortly after *the most popular hacker conference in the world,* and his immediate social circle was poised to take up opposition against Adobe almost instantaneously using a medium that would strike close to home and that was also embedded in hacker culture. Sklyarov's arrest struck at a key value in the hacker community: the free sharing of information. Although Adobe had enjoyed the support of the Internet community for some time, it now faced rebuke and boycott with protests organized at its San Francisco headquarters and other global offices just five days after the Sklyarov arrest. Because of activists' rapid advocacy,

drop the charges!
Dmitry Sklyarov,
a Russian student
and programmer,
is awaiting trial
in California on
spurious charges.
Help send him
back to his family.

Figure 4.3
Pamphlet about Dmitry Sklyarov and his wife and children distributed by the EFF, freedmitry.org, and freesklyarov.org.

Figure 4.4
T-shirt of a protester outside Adobe headquarters in San Jose, California, July 23, 2001.

the movement enjoyed a significant victory when twenty-four hours prior to protests (organized by the EFF, freedmitry.org, freesklyarov.org, and the Coalition to Free Dmitry), Adobe withdrew support for the FBI's handling of the AEBPR case.

Adobe tried to distance itself from Sklyarov's prosecution in an attempt to recapture some of the public-relations ground it had lost at the hands of activists, who painted Adobe as a bullying corporation. Adobe's reversal of position was sudden and stark. On the morning of July 23, 2001, it stated its position on its Web site: "Adobe fully supports the U.S. Government's decision to investigate the potential violation of U.S. copyright laws by ElcomSoft and has cooperated with their investigations. Adobe's goal is to help protect the copyrighted works of authors, artists, developers and publishers, and to stop the sale of this cracking software in the U.S." (quoted in Planet eBook 2001b).

Less than four hours later, after negotiations with the EFF and with protesters outside its doors, Adobe, in a joint statement with the EFF, turned its back on the government, stating, "We strongly support the DMCA and the enforcement of copyright protection of digital content. . . . However, the prosecution of this individual [Sklyarov] in this particular case is not conducive to the best interests of any of the parties involved or the industry" (quoted in Planet eBook 2001b). And so Adobe managed to stave off protests that had been organized by the EFF and other organizations.

Sklyarov and the Courts

Although Adobe's reversal can be claimed as a victory for the digital rights movement, the government proceeded with the prosecution regardless of Adobe's support, and it would take another year for the issue to be settled. The Sklyarov case was important to the US government and copyright owners because it was the first criminal prosecution under the DMCA's anticircumvention provisions. Under the DMCA, Sklyarov faced five years in prison and up to $500,000 in fines. Advocacy groups almost immediately framed the Sklyarov case as important for fair use and free speech and portrayed Sklyarov's and ElcomSoft's activities as helping consumers exercise these legal privileges. In a press release the day of Sklyarov's arrest, the EFF stated: "The Advanced eBook Processor appears to remove these usage restrictions, permitting an eBook consumer to enjoy the ability to move the electronic book between computers, make backup copies, and print. Many of these personal, non-commercial activities may constitute fair use under U.S. copyright law" (EFF 2001b).

Jennifer Granick, clinical director of the Stanford Law School Center for Internet and Society, likewise stated, "The DMCA says that companies can use technology to take away fair use, but programmers can't use technology to take fair use back. Now the government is spending taxpayer money putting people from other countries in jail to protect multinational corporate profits at the expense of free speech" (quoted in EFF 2001b).

Over the next few months, many movement actors became involved in Sklyarov's defense. The Web sites freedmitry.org and freesklyarov.org set up defense funds, developed mailing lists, distributed paraphernalia, and posted updates on Sklyarov's defense strategy. The EFF took a central role in coordinating protests, lobbying legislators, and meeting with representatives of the US Justice Department in attempts to get the government to drop the charges against Sklyarov. Although the EFF did not succeed in getting the government to drop the charges, it managed to bring significant media attention to the case. Lawrence Lessig, a well-respected legal scholar, an intellectual leader for the movement, EFF board member, and founder of his own movement strategy and organization,[2] published an editorial in the *New York Times* criticizing the DMCA (Lessig 2001). Even Congress weighed in when Representative Richard Boucher (D–VA) denounced the Justice Department's actions and proposed an amendment to the DMCA to exempt circumvention technologies that might help users exercise fair use (EFF 2001c).

When the Justice Department refused to free Sklyarov, the EFF used Attorney General Ashcroft's own words against him. When as a senator Aschroft had supported the DMCA, he had stated, "I think it is worth emphasizing that I could agree to support the bill's approach of outlawing certain devices because I was repeatedly assured the device prohibitions . . . are aimed at so-called 'black boxes' and not at legitimate consumer electronics and computer products that have substantial non-infringing uses" (quoted in EFF 2001c). The EFF published this quote in a press release subtitled "Boucher & Ashcroft Speak against Criminalization of Legitimate Software," implying that the attorney general was at odds with himself in his prosecution of Sklyarov.

Despite the EFF's heavy involvement in lobbying, raising public awareness, and coordinating protest and defense strategies, Sklyarov and ElcomSoft chose criminal attorney Joseph Burton to represent the case in court. They would need him. In late August 2001, Sklyarov and his employers at ElcomSoft were indicted on multiple counts of copyright violation and conspiracy and faced a potential $2 million fine and a twenty-five-year prison sentence. The indictment only served to further alienate the movement's constituency from the DMCA with continued protests and negative press. From the day of the indictment to the day of the arraignment, where Sklyarov and ElcomSoft pleaded not guilty, the Sklyarov case brought attention to the DMCA and to the digital rights movement.

The case itself did not go to court until December 2002, almost eighteen months after the initial arrest. By that time, Sklyarov had been allowed to return to Russia (after being held in the United States for five months) under a plea-bargain agreement where the government would use his testimony against his employer, ElcomSoft. Interestingly, both the government and ElcomSoft had Sklyarov testify on their side. In the twelve months preceding the case, ElcomSoft sought to get the case thrown out on a series of constitutional and jurisdiction issues. This strategy points to important ways in which the DMCA has come to be challenged through institutional settings such as the court system.

The first challenge involved the DMCA's jurisdiction, an issue that has been central in many discussions of the global nature of modern copyright and the Internet. Claiming that the alleged crimes were committed on the Internet, the defense argued that they were outside US jurisdiction. Although these arguments were continuously made during the early days of the Internet, they fell on deaf ears because courts have argued that even though the acts are questionably extraterritorial, the servers and the hardware are bounded within national territories that are subject to the laws of the land or international law. The issue of extraterritoriality and the reach of American law reappears often and has generally occupied intellectual-property scholars for the past fifteen years. The emergence of global governance structures such as the World Trade Organization and WIPO-administered treaties has put significant pressure on all governments to normalize their intellectual-property laws. This normalization has met with some measure of success, but also a great measure of resistance, particularly from developing nations who are hard pressed to see the advantages of technology transfer when duplication and pirating tend to be more immediately profitable.

Perhaps more important for the formation of the digital rights movement's framing of these issues was ElcomSoft's claim that the DMCA violated the US Constitution. Arguments regarding the constitutionality of the DMCA had been leveled at the statute since its inception, but ElcomSoft went further. Previous arguments citing First Amendment violation by the DMCA had been premised on issues of fair use, arguing that a restriction on fair use would be a restriction on a person's capacity to use citation and copying in excerpts for the purposes of free expression. ElcomSoft now argued that software code was a form of speech itself and that by banning certain types of code the DMCA was banning speech.[3]

In this vein, ElcomSoft made several points.[4] First, it noted that although the government is within its right to govern some speech, provisions that do so should withstand strict scrutiny on whether they are content neutral (which would be within the bounds of regulation) or *content specific* (whose regulation would be questionable). ElcomSoft argued that government regulation of computer programs in the DMCA was in fact constitutionally questionable because the regulation was content specific yet not narrowly tailored to suit government interests. It noted that "the anti-trafficking provisions seek to suppress computer code that indicates *how to circumvent technological* measures protecting copyright" (United States v. ElcomSoft, US District Court [ND Calif., San Jose Div., 2002], Memorandum of Points and Authorities in Support of Motion to Dismiss Based on First Amendment, emphasis added). The argument implies that the computer code is both a lock pick and instructions on how to pick the lock. Because it is those instructions to the computer on how to crack protected content that are illegal, the restriction on speech is very content specific.[5]

Second, according to ElcomSoft, the ban on code as speech substantially burdened speech more than was necessary to achieve the ends of copyright law. The key to this argument, mirroring Lessig's argument about the nature of architecture/technology and regulation, is the questioning of the connection between the means of regulation and the ends that regulation is attempting to achieve. Lessig calls this type of regulation "indirect regulation" (1999, 129). For example, a national identification system has been a hotly debated issue in our country and in Europe. In the United States, a national identification system would be difficult to implement because "right to privacy issues" and questionable constitutionality would create a base for heavy opposition,[6] so the government has skirted the potentially inflammatory measure by doing something altogether different. It has regulated not individuals, but the structures that individuals use in their everyday lives—for example, airline travel. By mandating airlines to require some form of official identification for passengers, the government has ensured that a great majority of its population will be compelled to acquire such identification and at the same time has circumvented the privacy issue. The regulation "everyone must have a state-issued identification" was achieved indirectly by the regulation "all who wish to fly on an airplane must provide official identification." Lessig questions the legitimacy of these types of tactics and wonders whether nesting the ends of a less popular regulation within the means of another regulation is a sleight of hand that is inherently undemocratic: "The point is not against indirect regulation generally. The point is instead about transparency. The state has no right to hide its agenda. In a constitutional democracy its regulations should be public. And thus, one issue raised by the practice of indirect regulation is the general issue of publicity. Should the state be permitted to use nontransparent means when transparent means are available?" (1999, 135). Lessig goes on to say that technology can function in a similar fashion. By regulating some technology that people use, the government may avoid directly regulating behavior that, for instance, the technology makes possible.

Third, ElcomSoft made a similar argument when questioning whether the DMCA could survive strict scrutiny. As noted earlier, content-specific regulation of speech is within the government's power if that regulation does not overburden free speech in general. ElcomSoft argued that what the government in fact was doing was indirectly regulating speech by attempting to regulate infringers. That is to say, under the guise of protecting copyright, the DMCA's anticircumvention provisions were in fact regulating a broad base of activities that fell outside the intended ends of the copyright statute.[7] By regulating circumvention technologies, the DMCA was in fact regulating activities that, although potentially detrimental to copyright owner's control over content, were nonetheless consistent with the spirit of copyright. ElcomSoft opined: "The AEBPR does not lead inexorably to the infringement of copyrights. For example, a blind man could use AEBPR in conjunction with a program to read eBooks aloud, because the technological measures in the Adobe eBook reader software must be circumvented so that the program can convert the text into an audio file. Such activity is plainly legal, not to mention beneficial" (United States v. ElcomSoft [2002], Transcript of Testimony).

And last, ElcomSoft argued that the DMCA placed undue burden on a third party's free-speech rights. The free speech of not only the designers of software, but also the users of content mediated by digital technologies such as eBook readers was affected. The DMCA did this in three important ways. First, by requiring a technological measure that indiscriminately guarded content, the DMCA was excluding users from access to works that were already in the public domain. Second, even if a work is not in the public domain, the copy-protection technologies would keep users from making fair uses. This particular point has been debated at length by others, not least the US Copyright Office (2001), whose view is that copyright owners are under no obligation to provide for fair-use access, that the law is under no obligation to compel them to do so, and that users have other means of making fair-use access, such as through non-digital or unprotected versions of the work (in other words, using a print book as opposed to an eBook). ElcomSoft claimed that the DMCA is unconstitutional because it is vague and exceeds the powers granted to Congress by the intellectual-property clause. Because the language of the statute notes that the intent of the designers of circumvention technology (whether primarily to infringe or not) is part of how a technology and its designer should be judged, the vagueness would produce a chilling affect among programmers and users of circumvention technologies. Furthermore, the fact that technological protection measures would protect content in perpetuity (regardless of whether the work is still protected by copyright or has passed into the public domain) was noted as a reason why the DMCA went beyond the limiting stipulations of the intellectual-property clause.

ElcomSoft's motions were supported in briefs filed by various organizations associated with the movement.[8] The briefs were indicative of the framing of the issue that organizations in the movement wanted to establish. They attempted to tie fair use and the limitation on copyright owners' exclusive rights to protection under the First Amendment and a perceived plasticity for the doctrine explicitly noted by Congress. They therefore argued that the government's broad application of the DMCA had negative effects on free speech by way of constricting fair use. Copy-protection technology was illustrated as a force that in effect limited fair use by protecting content that (1) may be part of the public domain or (2) should be accessible for fair uses even if copyrighted. The briefs understood technological enforcement as standing in the way of fair use in digital media and noted that the DMCA, by protecting technological enforcement measures, also stood in the way of fair use. As shown in subsequent sections in this chapter, the court did not agree with this view because it saw fair-use access to digital media as a matter of convenience. Users, the court noted, could still have fair use "the old-fashioned way," such as copying a quote by hand.

The movement's and the government/copyright owners' viewpoints on this issue were ultimately incommensurable. The movement saw consumption of content as mediated primarily through digital technology. Therefore, it argued, preserving the rights of the analog world in the digital world is of great importance. The government did not see consumption of content as mediated solely through digital technology, so, according to it, access to a digital work is a matter of convenience, especially if the work can also be accessed via a printed book or a tape recording. For this reason, the court did not see why it should view the DMCA as blocking fair use or free speech. So long as content is available in other formats, why should it matter that a customer cannot have the convenience of the copy/paste function in a digital book? But that was the movement's point: to bring the privileges enjoyed in analog media to digital media.

The court rejected all of ElcomSoft's motions to dismiss the case. Of interest to the movement was the irreconcilable understandings of speech as expressed by ElcomSoft and the court. The court argued that ElcomSoft's claims that the DMCA is unconstitutional because it regulates code, which is a form of speech, was not convincing because code has both "speech" and "nonspeech" elements. Given such tight coupling, the government, if pursuing legitimate interests (such as preserving copyright), may infringe somewhat on speech.[9] The key to the court's understanding of the issue is a perceived dichotomy between code's speech and nonspeech elements and the idea that what is to be protected is expression.[10] Thus, by understanding the DMCA as regulating the actions of code as opposed to the expression of code, the court ruled that the DMCA was not running afoul of free speech.

The court's attempt to distinguish code's nonspeech and speech elements is both arbitrary and based on a problematic understanding of what code actually is. Code, a computer program, whether as 1s and 0s or source code, is a set of instructions to a computer on how to perform a specific task. By the statement "When speech and non-speech elements are combined in a single course of conduct, a sufficiently important government interest in regulating the non-speech element can justify incidental intrusions on First Amendment freedoms" (United States v. ElcomSoft [2002], Indictment), the court implied that in a given program there are both speech and nonspeech elements. This is wholly incorrect. The whole of the program is an instance of speech—perhaps not expressive, but speech nonetheless.

An analogy may help to make this point clear. Compare instructions written for people on how to reverse engineer a combination lock and instructions written for a computer (that is, code) for how to break encryption systems. Code is a set of instructions for a computer; instructions on how to reverse engineer a lock are instructions for a human. Regardless of the audience, human or computer, the instructions are still just that—instructions. Yet what is the rationale for qualifying one as wholly a speech element and the other (code) as composed of speech and nonspeech elements? The court implied that code results in the action of breaking encryption, whereas instructions on how to reverse engineer a lock does not necessarily result in the breaking of the lock. This conclusion is spurious. The only reason that code might result in the breaking of encryption is that some person loaded the code into a computer and executed it, just as the only reason a lock might be broken is that some person "loaded" the instructions on how to break a lock into his brain and executed them. The point is that the code itself has no magical functional elements. It does nothing itself; rather, it is the computer following the instruction the code has given it that does something, and it is the person who loads that code who is ultimately executing it, whether it breaks a combination lock or encryption. The claim that there is some actually functional nonspeech element to code is incorrect.

If there are no nonspeech elements to code, then the claim that "[w]hen speech and non-speech elements are combined in a single course of conduct, a sufficiently important government interest in regulating the non-speech element can justify incidental intrusions on First Amendment freedoms" is invalidated. What, then, remains as a rationale for making a distinction between instructions for computers and instructions for people?

It appears that the court and the law have made the arbitrary distinction based on ease of execution. Instructions to a computer on how to break encryption are much more easily executed than instructions to a person on how to break a lock. All a person has to do is load the code on the computer. Conversely, breaking a lock incurs a greater cost on the breaker; it requires materials, and the breaker of the lock can be more easily caught than the "pirate" using code to break encryption. Thus, it seems that because code is so easy to execute in contrast to instructions on how to break a lock, the court created categories of speech versus nonspeech to justify regulating one instance of speech and not another. This misconception played itself out in the DeCSS case as well, where the courts effectively banned the source code from the Internet not because it had some nonspeech element that the court could control (activists made plain that it did not), but because those instructions could easily be used.

Indeed, hackers refer to programs as "information": the source code for a program is literally a step-by-step manual on how a computer should accomplish a given task. One can learn much about programming from those instructions, and this principle is basic to open-source development, for example. The idea that there are no nonspeech elements in code is the reason why ElcomSoft argued that the DMCA was in fact controlling speech when controlling code and as such was overburdening free speech.

The court was on more solid ground when it argued that, notwithstanding arguments about whether code is or is not speech, its ban was not directed at what code says, but rather at what it can lead to. The court in that sense admitted that the DMCA is trying to regulate the conduct that code makes possible in lieu of being able effectively to regulate infringers directly. However, because the DMCA is not an instance of the government's disagreeing with an opinion expressed in code, the control of code is permissible. But this view is premised on the idea that code is not expressive of a political opinion. What if code were written both as instruction and as political statement? Interestingly, the process of making code a more political statement is under way in the digital rights movement, and it underlies the idea of technological resistance. As examples given later on illustrate, some programs are written with express political purpose (see the discussion of iTunes hacks in chapter 7). In such cases, the expressive function of software (what it says about what you believe in) is based on what activists and designers of the software say it means. In that sense, the meaning of the software depends on assignations that are context and user/viewer specific. Thus, although activists say that designing and using a specific program or code can mean the person who designs or uses it supports the digital rights movement, the government can challenge this interpretation by saying that this connection means nothing at all. But what if the expressive political statement were built into the code: Can the government still suppress it based on an arbitrary distinction between speech and nonspeech elements of code? The DeCSS case makes the implications of this question concrete. When the DeCSS source code was banned from publication, copies of it immediately showed up all over the Internet in myriad forms, as in the following verse, which embeds the code in political and ideological statements about the nature of copyright:

Now help me, Muse, for
I wish to tell a piece of
controversial math,
for which the lawyers
of DVD CCA
don't forbear to sue:
that they alone should
know or have the right to teach
these skills and these rules
CSS is
no exception to this rule.
Sing, Muse, decryption
once secret, as all
knowledge, once unknown: how to
decrypt DVDs.
Arrays' elements
start with zero and count up
from there, don't forget!
Integers are four
bytes long, or thirty-two bits,
which is the same thing.
To decode these discs,
you need a master key, as
hardware vendors get. (Touretzky n.d.)

The distinction between the speech and nonspeech elements is a slippery slope that hackers and activists will surely exploit. As the poem shows, it is quite easy to embed what courts would try to declare is a nonspeech element of software in clearly contentious speech.

The court also rejected claims made by ElcomSoft regarding the overbroad effect the DMCA would have on the free speech of third parties—those not writing the code, but using eBooks, for example. "First, the DMCA does not 'eliminate' fair use. Although certain fair uses may become more difficult, no fair use has been prohibited. Lawful possessors of copyrighted works may continue to engage in each and every fair use authorized by law. It may, however, have become more difficult for such uses to occur with regard to technologically protected digital works, but the fair uses themselves have not been eliminated or prohibited" (United States v. ElcomSoft [2002], Motion to Dismiss on Constitutional Grounds). This statement by the court is emblematic of the government's continued understanding of how users should engage in digital media. By viewing access to and use of digital technology as a matter of convenience, the court rationalized that so long as the user can have fair use the old-fashioned way (by copying by hand, for example), the DMCA was not really burdening them any further. Also, the court reasoned that even if a copy-protection technology prevents users from making fair uses of works in the public domain, it is not the same as granting the publisher exclusive rights because, again, the user can have full access to a work through other means or can copy it by using nondigital methods—that is, writing it out by hand.

The court's understanding of how a user should access digital content denied the technological convenience of digital technology to users. Also, the court found that fair use cannot be equated with free speech because although there may be some references in precedent to the need for fair use to facilitate free speech, the issue was unsettled, and no court-established or statutory right exists for it. The court missed the importance of technological enforcement. It failed to understand how the digital rights movement was framing fair use: as crucial to free speech and as an expansion of privileges held in the real world into the virtual world.

On December 2, 2002, the case United States v. ElcomSoft Ltd. went to federal court in San Jose. The case played itself out uneventfully, and in the course of testimony Adobe representatives noted that even though they had looked throughout the Internet, they had not been able to find a single instance of a protected eBook being made available in free-distribution networks. Sklyarov and ElcomSoft came across not as nefarious criminals, but as enterprising employee and business that had been caught in the intricacies of US copyright law that was strangely being applied to supposed crimes committed outside of American soil.

In one pointed exchange between Sklyarov and the government prosecutor, Scott Frewing, Sklyarov pointed out that Russian law made illegal the license agreements that deny users the right to reverse engineer. When asked if he violated the license agreement, Sklyarov noted, "I don't think so. . . . [I]n fact, as far as I know, according to Russian law, I have [the] right to reverse engineer any software for purposes of compatibility and if license conflict [*sic*] with Russian law, Russian law has priority" (United States v. ElcomSoft [2002], Transcript of Testimony).

In the course of the trial, ElcomSoft's attorney also tried to raise the issue of fair use by suggesting that Adobe, in fact, was preventing users from exercising their rights over copyright material, but the court refused to allow that line of questions, and only during Sklyarov's testimony did it become clear that the software had clear potential to be used for purposes that were well within legal bounds.

In the end, the jury found ElcomSoft not guilty on all counts, and the DMCA suffered a significant defeat as a law that can be enforced via criminal prosecution. The government has not attempted to try another criminal case under the DMCA since the Sklyarov victory, choosing instead to use provisions in other sections of the copyright statute to bring charges of contributory infringement on technology manufacturers and users. For their part, content owners continue to rely on the DMCA as a tool to ground their lawsuits against individuals distributing and using circumvention technologies, a strategy that has proven very successful.

Conclusion

The Sklyarov case remains important in the history of the digital rights movement because it helped hackers, the EFF, and other activist networks capture the issues of fair use and freedom of speech for movement activism and gave the DMCA some very negative press. Just a short month following Sklyarov's arrest, the idea of prosecuting him appeared grossly unjust, and support for the DMCA and Adobe sagged. At the same time, the movement was enjoying the positive glow of championing free speech and fair use in digital media, and the pejorative image of hackers and movement activists as pirates and criminals did not stick. In August 2001, a *Washington Post* editorial noted,

> [Sklyarov's arrest] is . . . one of the most oppressive uses of the law [DMCA] to date—one that shows the need to revisit the rules Congress created to prevent the theft of intellectual property using electronic media. . . . Programs to break copy protection schemes can be used to facilitate fair use, as well as infringing uses of copyrighted material. Simply banning the dissemination of such programs, without reference to the purpose of the dissemination, inhibits the use of intellectual property far more broadly than does the copyright law itself. ("Jailed under a Bad Law" 2001)

The Sklyarov case propelled activists into action, helping the movement articulate how the DMCA and technological enforcement were affecting free speech. (See chapters 5 and 6 for a discussion of the DeCSS case, which also helped articulate how the DMCA affects fair use, free speech, and innovation.) The Sklyarov case was reported by global media outlets and elicited scorn of the government's actions even from those outside the immediate circle of hackers and technologists who were affected. It also helped establish the EFF as an important organization working on issues of digital rights. Although the EFF had been working on these issues since at least 1990, by taking on Adobe it established itself as a significant force able to win against formidable opponents.

Today one can easily download a newer version of AEBPR online. The media attention and the opportunity to challenge the US government and Adobe gave the movement a tangible issue to fight and a sympathetic figure with whom those previously outside the movement's orbit could identify. The movement itself ably used digital networks to deploy sympathetic images of Sklyarov with his family and children, showing him not as some evil "hacker" being rightly prosecuted, but as an educated, family-oriented, shy-looking young man who had exercised academic freedom by presenting his findings on eBook security and was now under arrest.

5 DeCSS: Origins and the Bunner Case

Like the Sklyarov case, the DeCSS case, another important landmark in the evolution of the digital rights movement, involves the development of a technology that infringed on the DMCA's anticircumvention provisions. The movement framed the prosecution of individuals who were linking to the DeCSS source code as an infringement on free speech and a disincentive to innovation. This chapter shows the changing configuration of technology and users. Whereas the AEPBR was a consumer product that came to be politicized through the legal process and movement framing, DeCSS, which emerged from hacker groups rooted in the open-source/free-software movement, came with some politics already articulated in its design. DeCSS designers imagined users who had advanced technical capabilities but also saw them as key intermediates who would bring DVD players to wider publics using the Linux operating system. DeCSS served as a "locale" for projecting other important movement frames and highlighting the logical problems in the way the law understood journalism, code as speech, and technological innovation. Like the AEPBR before it and iTunes after it, the DeCSS technology became meaningful beyond its functionality, and the discourse surrounding it helped solidify important movement beliefs.

Although both this chapter and the next focus on DeCSS, they examine different cases that highlight the different approaches that the content industry and movement advocates took in arguing against or for DeCSS's legitimacy.

DeCSS

The DeCSS story has its beginnings with the Content Scrambling System (CSS) present in all DVDs carrying commercial films since 1999. Matsushita, the parent company for Panasonic, and Toshiba jointly developed the CSS

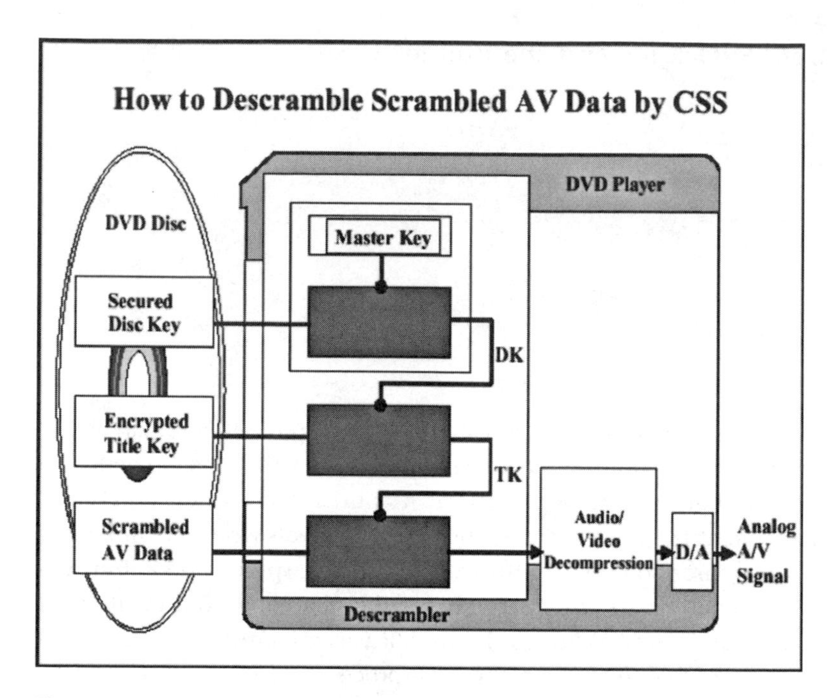

Figure 5.1
Schematic of how CSS works. From Touretzky n.d.

technological protection measure and incorporated it into the standard DVD format. The CSS technological protection measure uses a series of keys to encode the video content on DVDs and establishes the technological system that enforces the DVD's licensing terms. As the schematic in figure 5.1 shows, there are three encryption keys associated with DVD playback: (1) the master key on the DVD player; (2) the disk key on the DVD; and (3) the title key, also on the DVD. A DVD player uses its master key to decrypt the disk key, which it can then use to decrypt the title key. In turn, the title key can be used to decrypt the actual content on the DVD.

Because all DVDs are encoded with CSS, all DVD players must have a licensed CSS master key that allows them to decrypt the content. Because the master keys are licensed, all DVD player manufacturers pay a royalty to the movie industry through its representative, the DVD Copy Control Association (DVD CCA). It is important to note that CSS is not considered a DRM system because the user is given no privileges over the content on the DVD. In contrast, with a DRM system a user might have some limited copying or distribution privileges. Thus, CSS is not like the eBook copy-protection technology.

As noted earlier, the CSS does enforce licensing terms that (1) dictate regional playback permissions so that DVD players in the United States will not play DVDs bought in China or Europe; (2) designate certain sectors on the DVD as non-fast-forward sectors, such as the parts on the DVD that contain the standard FBI warning and, in some cases, commercials and adverts; and (3) prevent content on DVDs from being copied on VHS recorders by interfacing with the Macrovision technological protection measure, which is standard on VCRs.[1]

As personal computers (PCs) and then DVD players became mainstream consumer devices in the late 1990s, DVD players became standard hardware on PCs. However, all DVD players made for PCs at the time had software support for only Windows systems, so DVD players for alternative operating systems, such as the open-source system Linux, were left out of this market. Open-source systems must have "open" applications that make source code available, so these players and the software that ran them could not comply with the licensing terms. To get around this hurdle, the open-source community set out to design drivers and other software that would allow the Linux system to use DVD players.

In October 1999, an anonymous German programmer known only by his online name "Ham" cracked the CSS encryption algorithm and released it to two hacker groups working on DVD player applications for the Linux operating system: the Drink or Die (DoD) group from Russia and the Masters of Reverse Engineering (MoRe) group, whose members were distributed throughout Europe. Both groups used the decryption algorithm to design applications that would not only read content from DVDs but also "rip" the content from the DVD and allow it to be stored in an unscrambled format on a PC's hard drive. DoD designed the application DoD DVD Speed Ripper, and MoRe designed DeCSS. Although the two applications were designed simultaneously and released within weeks of each other, DeCSS garnered the majority of media attention for two reasons. First, DeCSS worked for every movie title in DVD format available at the time; in contrast, Speed Ripper had some difficulty with certain titles. Second and most important, the DeCSS source code was released to the DVD Linux development community, making the CSS decryption algorithm a matter of public knowledge. This meant that anyone could look at the DeCSS source code and glean from it how CSS scrambled video content. Therefore, anyone with enough knowledge of cryptography could design his or her own DeCSS-type program.

The significance of having the DeCSS source code available to the public cannot be overstated. It impelled the DVD CCA to start its legal campaign against individuals and organizations linking to or distributing the source code. It was this legal campaign that gave the digital rights movement another opportunity to frame its ideals along lines congruent with free speech and fair use.

The Bunner Case

Two months after DeCSS was released on the Internet, the DVD CCA and the MPAA separately filed suits in California and New York against a host of individuals posting or linking to posts of the DeCSS application and its source code. The two cases took different approaches in explaining why DeCSS posed a danger to the motion picture industry.

In DVD CCA, Inc. v. Andrew Bunner et al. (California Superior Court, Santa Clara County, 1999), the DVD CCA sought an injunction on distributors and Web sites linking to the DeCSS source code and application, arguing that posting DeCSS was a misappropriation of trade secrets. In its complaint to the court, the DVD CAA argued that only licensed DVD player manufacturers and their affiliates had access to the CSS code and that those manufacturers were bound by the license agreement not to reveal the CSS code. The existence of DeCSS implied that CSS must have been accessed in breach of the license agreement, so those individuals posting DeCSS, because it was developed using a misappropriated trade secret, must be enjoined from continuing to post the information. The DVD CCA named twenty-five defendants and more than five hundred "John Does" in its complaint to the California Superior Court in Santa Clara County. Some of the named defendants were from Denmark, France, England, Germany, and Norway. The DVD CCA argued that hackers had illegally obtained master keys by hacking the well-known DVD player application X-ing for Windows systems. They noted that anyone running X-ing on their computers would have to click-through a license agreement that precluded the user from reverse engineering the software.[2]

Many of the defendants in the case sought legal advice and help from the EFF, and so it once again found itself as a central organization in voicing important points about the DMCA's impact and the growing technological enforcement of copyright. The EFF noted that to enjoin hackers, cryptographers, research scientists, and Linux developers from posting and distributing the source for DeCSS would be an abridgement of First Amendment rights. Furthermore, the EFF noted that a court-administered gag order on discussing these types of encryption technologies would have a chilling effect on research and development in cryptography.

YBP Library Services

POSTIGO, HECTOR.

DIGITAL RIGHTS MOVEMENT: THE ROLE OF TECHNOLOGY
IN SUBVERTING DIGITAL COPYRIGHT.
Cloth 244 P.

CAMBRIDGE: MIT PRESS, 2012
SER: INFORMATION SOCIETY SERIES.

AUTH: TEMPLE UNIV.

LCCN 2012-4559
ISBN 0262017954 **Library PO#** FIRM ORDERS

8395 NATIONAL UNIVERSITY LIBRAR
App. Date 5/07/14 SOC-SCI 8214-08

List 32.00 USD
Disc 14.0%
Net 27.52 USD

SUBJ: 1. COPYRIGHT & ELECTRONIC DATA PROCESSING.
2. HACKTIVISM.

CLASS K1447.95 DEWEY# 345.02662 LEVEL ADV-AC

YBP Library Services

POSTIGO, HECTOR.

DIGITAL RIGHTS MOVEMENT: THE ROLE OF TECHNOLOGY
IN SUBVERTING DIGITAL COPYRIGHT.
Cloth 244 P.

CAMBRIDGE: MIT PRESS, 2012
SER: INFORMATION SOCIETY SERIES.

AUTH: TEMPLE UNIV.

LCCN 2012-4559
ISBN 0262017954 **Library PO#** FIRM ORDERS

8395 NATIONAL UNIVERSITY LIBRAR
App. Date 5/07/14 SOC-SCI 8214-08

List 32.00 USD
Disc 14.0%
Net 27.52 USD

SUBJ: 1. COPYRIGHT & ELECTRONIC DATA PROCESSING.
2. HACKTIVISM.

CLASS K1447.95 DEWEY# 345.02662 LEVEL ADV-AC

In their statements to the court, the defendants, some of whom were well-known cryptographers, noted that CSS was highly flawed and highly susceptible to attack and that DeCSS was consistent with professional practices of exposing security flaws in supposed trusted systems. One defendant, a graduate student in computer science at Berkeley, said: "I examined the CSS encryption algorithm soon after its flaws were first revealed to the public. In my opinion, the CSS was extremely poorly designed. . . . I believe breaking it would make a fine homework exercise for a university-level class in cryptography and code breaking" (DVD CCA v. Bunner et al. [1999], Declaration of David Wagner).

The defendants also argued that DeCSS would be of great help in designing software to play DVDs on the Linux operating system, a system ignored by the DVD CCA. They defended their decision to keep the code public because they "felt that providing others with access to the DeCSS program, and thereby enabling Linux users to play DVDs, was important because it would make Linux more attractive and viable to consumers, thereby making Linux a more viable and accepted Operating System platform" (DVD CCA v. Bunner et al. [1999], Defendant Andrew Bunner's Declaration).

Furthermore, they pointed out that claims that DeCSS would lead to rampant piracy were greatly exaggerated, noting that most computer and Internet users lacked the technological know-how and communication resources to traffic in large volumes of media. For example, one programmer explained that hard-drive limitations alone would preclude computer users from pirating movies. Hard drives had about a thirty-gigabyte capacity at the time, enough for only three or four movies. Furthermore, it would take days to properly encode the files to be viewed on the computer screen, and no commercial DVD burners were yet available. "CSS primarily prevents one from building DVD players without permission from the DVD industry, and does not prevent large-scale copy of DVD content," noted one defendant (DVD CCA v. Bunner et al. [1999], Declaration of David Wagner). The majority of DVD piracy, he continued, occurs through large operations where the content of DVDs is imaged directly onto blank DVDs, encryption included.

Defendants from outside the United States also submitted statements for the court. Most interesting for our purposes are the statements from defendants from Norway, the home country of Jon Johansen, the young man partially credited for designing and posting DeCSS. Although Johansen was not named a defendant because he initially removed his posting of DeCSS, many defended the posting of DeCSS in Norway, citing

Norwegian statute that gave users the right to reverse engineer. They argued that this right could not be given up with a click-through agreement administered by X-ing. Although the merits of a click-through license have not been settled in the United States or Norway, the conflict between local copyright laws and international copyright regimes is a recurring theme in the global regulation of intellectual property. It was highlighted when many supporters of DeCSS questioned how the United States justified enforcing its laws (on trade secret, for example) on citizens of other countries.[3]

Much of the frustration expressed by the defendants centered on what appeared to be unfair regulation of regional encoding that prevented DVDs made in the United States from being viewed in Asia or vice versa. Defendants complained that CSS was not designed primarily to protect content on DVDs because a weak encryption would appear vulnerable to even the novice cryptographer. Rather, defendants believed that the CSS was a way to control traffic of movies from one part of the world to another so that the film industry could control release dates and prices for different regions.[4] They saw this control as an unfair business practice and thus felt justified in circumventing the CSS.

Despite the arguments made by the EFF and the defendants, Judge William Elving of the California Superior Court decided to grant a temporary injunction on distribution of the DeCSS code and application. At the same time, however, he refused to enjoin defendants from linking to the code because, he reasoned, that linking was a central function of the Internet. The judge noted that although there was no clear indication that the defendants or even Jon Johansen had violated a click-through license in order to develop DeCSS, the circumstantial evidence overwhelmingly suggested that the code had been derived by some violation of the license. This issue was and continues to be debated. Many of the hackers who had commented on DeCSS during its development and release noted that the makers of X-ing had failed to encrypt their keys in the program itself and therefore had made it very easy for the keys to be found. The makers of X-ing denied having exposed the keys in this fashion, knowing well enough that such a breach of security on their part would also make them liable for the loss of trade secret. Even if the keys were unencrypted, they said, hackers had to look at the X-ing's source code and therefore must have violated X-ing's license agreement.

In one of his final remarks on the case, Judge Elving noted that although the code was barred from the Internet, discussion of it was not. He noted, "Nothing in this Order shall prohibit discussion, comment or criticism, so long as the proprietary information identified above is not disclosed or distributed" (DVD CCA v. Bunner et al. [1999]). Defendants perceived this directive as inconsistent—the equivalent of saying you can discuss *War and Peace*, but no one should be able to read it. This difficulty did not go unnoticed at the time, and the EFF appealed the injunction. Despite the setback, fifty-two of the seventy-two named defendants and most of the five hundred unnamed defendants continued to post DeCSS in some form or another.[5]

When the injunction was appealed, however, the Appeals Court of Southern California overturned it based on its interpretation of copyright law, trade-secret law, and precedents in First Amendment decisions concerning software as expressive or "pure" speech.

First, the appellate court found that the precedents that the plaintiffs had used to show that injunctions had been awarded in trade-secret cases did not compare to the case before the court. The DVD CCA had used precedents describing injunctions to direct violators of secrecy agreements, and the court reasoned that the Internet posters were not direct violators. Furthermore, the court found that because the DVD CCA was trying to bring suit against Andrew Bunner under trade-secret law, which is designed specifically for those who have voluntarily entered into a contractual agreement (something Bunner did not do), that law could come into conflict with the First Amendment. In such a case, protection of free speech would take precedence. *The court implied that had the plaintiffs couched their arguments under copyright law, then the court would have had to balance carefully between two constitutionally protected rights.* As it was, the appellate court found that software *is* a form of speech and that the lower court exercised prior restraint of the defendant's speech, something that has been consistently opposed by the Supreme Court (see, for instance, Universal City Studios, Inc. v. Corley, 273 F. 3d 429, US Court of Appeals [2nd Cir. 2001]). Importantly, the content industry argued both the Sklyarov case and the Reimerdes case, discussed later in this chapter, as copyright cases, so the court in those cases had constitutional grounding when it found for the content industry, a luxury the Bunner court did not have. Furthermore, the court of appeals in Bunner, unlike the court in Sklyarov, noted that although source code has an only functional speech element, that element should not preclude it from being protected as pure speech.

The issue of code's speech and nonspeech elements was an important theme in these early cases in the digital rights movement. As noted in chapter 4's analysis of the Sklyarov case, the court presented an arbitrary distinction between speech and nonspeech elements of code. As I argued in that chapter, the speech/nonspeech distinction is made for the sole purpose of regulating speech that has become commercially important but that is increasingly politicized. The central error in all understandings of code is that it is perceived as both expressive and functional. Code is not functional in any sense; it is a set of instructions to a computer. The computer is the functional component. *If code is written with the expressed intent of resistance, then it is expressive speech telling computers exactly how to help the movement achieve its goals.*

Not satisfied with the appeals court decision, the DVD CCA petitioned the California Supreme Court for further review. In its petition, it relied heavily on *Universal City Studios, Inc. v. Reimerdes* (111 F. Supp. 2d 294 [SD NY 2000]), which had just been decided in the State of New York. The New York court rejected as gross misinterpretation the view that software code is speech. In the DeCSS case, returning to the idea of speech versus nonspeech elements, the DVD CCA attacked the appeals court's refusal to see that DeCSS was primarily functional speech, which would mean that it required only an intermediate level of free-speech protection. It noted,

> By posting DeCSS on his website, knowing that it contained stolen trade secrets, Respondent Bunner engaged in no expressive discourse about issues of public concern. . . . The Court of Appeal erroneously applied the First Amendment doctrine against prior restraints that can be found in cases involving pure, political speech, such as New York Times Co. v. United States (1971) 403 U.S. 713 (the Pentagon Papers case) . . . without considering how dramatically different the speech in those cases was from the dissemination of stolen trade secrets here. The speech sought to be enjoined in *New York Times* . . . lay at the very heart of First Amendment concern—public debate about policy issues. (DVD CCA v. Bunner et al. [2001], California Supreme Court, Petition for Review).

But what constitutes public discourse of social concern and importance to debate about public policy? Surely, as code becomes technological resistance, it constitutes some form of speech about the way technologies ought to work. And is it not important discourse within communities of hackers? As these technologies become widely used, it is important for groups outside of those narrowly interested in the workings of technology to discuss its wider implications. The point is that as technologies are further marginalized by law yet remain valid forms of discourse and acceptable tools for getting something done, then they acquire meanings that are consistent with resistance. DeCSS, eBook, and later on iTunes hacks follow this trend from controversial consumer item to technological resistance. The point the DVD CCA made regarding this issue turns a blind eye to that reality.

The DVD CCA continued to see technologies such as DeCSS as "hackerware" and so took the case to the California Supreme Court, which remanded the appellate decision back for review. In its discussion of the merits of the analysis, the California Supreme Court disagreed with the appeals court interpretation of code as speech worthy of strict protection, choosing not to engage in an analysis of speech versus nonspeech elements, but rather focusing on whether trade-secret law allows for prior restraint. The court found that the injunction was content neutral (not concerned with the opinions expressed by the code, but rather by the code itself) and found that an injunction on Bunner was subject to intermediate speech protection. Finding that although a ban on posting impinged on Bunner's free speech, it served a legitimate government interest, and the court noted that the appeals court had wrongfully judged the DeCSS case as a case of prior restraint subject to the strictest free-speech analysis. The court did, however, request that the appeals court review the Superior Court's application of the injunction with respect to California's trade-secret law, inquiring whether CSS had entered into the public domain and no longer deserved trade-secret protection.

Although all the legal considerations in the case (e.g., trade-secret law) are beyond the scope of this discussion, what is interesting is how the EFF, whose lawyers argued the case, chose to interpret this decision. The California Supreme Court's argument that code merits only intermediate speech protection was a blow to the movement's claim that code ought to be protected speech in the highest sense. This argument was consistent with other decisions occurring at the time (such as in the Reimerdes, Corley, and Sklyarov cases) and should have been considered a defeat in the movement's attempt to define circumvention technologies as matters of free speech. The EFF, however, spun the decision in the best possible light, stating in a press release curiously titled "California Supreme Court Upholds Free Speech in DVD Case" that "the appeals court can now examine the movie industry's fiction that DeCSS is still a secret and that a publication ban is necessary to keep the information secret. . . . DeCSS is obviously not a trade secret since it's available on thousands of websites, T-shirts, neckties, and other media worldwide" (Cohn 2003).

The DVD CCA anticipated this response and quickly tried to have the case dismissed in the appeals court before the court could judge that CSS was no longer a trade secret. Cognizant of this tactic and, in fact, aware that DVD CCA had delayed judgment for more than two years under the pretext of clarifying the issue, the court denied the dismissal and reversed the injunction on Bunner on February 27, 2004, four years after the injunction had been issued. The court noted some important developments in the case. First,

> [t]he lawsuit outraged many people in the computer programming community. A campaign of civil disobedience arose by which its proponents tried to spread the DeCSS code as widely as possible before trial. Some of the defendants simply refused to take their postings down. Some people appeared at the courthouse on December 28, 1999 to pass out diskettes and written fliers that supposedly contained the DeCSS code. They made and distributed tee shirts with parts of the code printed on the back. There were even contests encouraging people to submit ideas about how to disseminate the information as widely as possible. (DVD CCA v. Bunner et al., Court of Appeal, State of California (6th District, 2004], Decision on Remand)

Second, the court noted that just because someone chooses to publish a trade secret does not mean that it ceases to be a trade secret or that the person is not liable for damages. Rather, once that information is in the public domain, no state action can bring it back without abridging speech. The court noted that the evidence demonstrated that in the Bunner case "the initial publication was quickly and widely republished to an eager audience so that DeCSS and the trade secrets it contained rapidly became available to anyone interested in obtaining them" and that "DeCSS had been so widely distributed that the CSS technology may have lost its trade secret status" (DVD CCA v. Bunner et al. [2004], Decision on Remand). Thus, because CSS had lost trade-secret status by the time the injunction was issued, the appeals court found that the Superior Court's injunction on Bunner overextended its powers under California's trade-secret act.

Conclusion

The outcome of the Bunner case was important for the digital rights movement in many respects. On the one hand, it was a victory for the movement. The EFF, as it continued to define itself as *the* SMO concerned with digital rights issues, showed that the content industry could be defeated in its attempts to censor technology that can potentially benefit consumers. The Bunner case (and really the whole of the DeCSS legal history) highlighted the potential of seeing code as speech and the consequences for the movement. It illustrated that technologies meant to allow user-centered functions, such as copying for personal backups or reverse engineering to design a DVD player, not only had a functional purpose but were meaningful as acts of political speech. On the other hand, the fact that the court failed to agree with the "code as speech" argument was a failure for the movement in that this particular frame did not gain traction in institutional settings. Therefore, although the movement was increasingly able to articulate its viewpoint along important themes (such as fair use and free speech), these themes and their related arguments continued to have a hard time finding positive reception outside the movement and its allies. In the Bunner case, the injunction was ultimately denied based on practical issues: the DeCSS had made the CSS code a matter of public knowledge, and it was too late to put the cat back in the bag. The decision was not based on perhaps the more important and enduring issues of fair use and free speech, which could have greatly helped the movement's cause in future court cases.

6 DeCSS Continued: The Hacker Ethic and the Reimerdes Case

The Reimerdes case, like the Bunner case, is important in the DeCSS history in that it showcases important frames for the movement. Furthermore, it is particularly illustrative of the dissonance between hacker practices, academic freedom, and user concepts of freedom of speech and legal/industry interpretations of those concepts. It shows how wider publics such as academics in the fields of cryptology and engineering can be drawn into a debate that starts off as a question of copyright policy. In that sense, the case demonstrates the power of movement frames to touch on the everyday technological practices of users with technological expertise and transcend into professional practices such as online journalism.

The Fight for a Preliminary Injunction

On January 15, 2000, a little more than two weeks after the DVD CCA filed suit against Andrew Bunner and others in California, the MPAA filed for an injunction on three individuals in New York State: (1) Shawn Reimerdes, who operated a Web site called dvd-copy.com, distributed the DeCSS program, and linked to other sites that also distributed the application; (2) Roman Kazan, who operated the Web site krackdown.com/decss, which also distributed the DeCSS source and application; and (3) Eric Corley, perhaps the most important for our discussion, who operated the *2600: The Hacker Quarterly* Web site. The print publication and Web site for *2600* magazine are edited by Eric Corley under the pseudonym "Emmanuel Goldstein" (the name of a character in George Orwell's *1984*).

Founded in 1984, the magazine is at the center of hacker subculture. The name of the publication itself is a reference to one of the most infamous instances in computer hacking known as "phreaking." *Phreaking* is a hacker term melding the words *phone* and *freak* and used to describe hackers who study and exploit weaknesses in telephone communication architectures. Phreakers are generally considered to be a subgroup within the broader hacker community and are usually minors younger than seventeen. The legal ramifications for hacking into the telephone system can be quite severe, and most leave the practice when the costs become potentially too high. In the 1960s, hackers discovered that a telephone user could access the operator mode in telephones and make free long-distance telephone calls if they whistled a tone with a frequency of 2600 hertz into the receiver of any telephone. The magazine title is thus a direct reference to one of the earliest of hacker practices, and when Universal named *2600*'s Eric Corley as a defendant in its lawsuit, it attacked a mainstay in hacker culture. Corley had also been named in the DVD CCA case, but in this instance he took a central role.

In both the Bunner and Reimerdes cases, the movie industry wanted to enjoin the defendants from distributing and linking to the DeCSS application and source code, but in the Reimerdes case the MPAA used the DMCA and copyright law as its basis for enjoining the defendants. Learning from the DVD CCA case, in which by this time the DVD CCA had been denied a temporary injunction by the California Superior Court, the MPAA presented its case from the perspective of the constitutionally important intellectual-property clause.

In its arguments for granting an injunction, the MPAA was affected by earlier failures in California. When the Superior Court chose to deny the DVD CCA a preliminary injunction in Bunner, DeCSS proponents interpreted this decision as an affirmation of their free-speech rights and so increased their efforts to distribute the code as widely as possible. Corley had actually been trying to do so ever since the code had been released in November 1999 (a month before he was named in either the Bunner case or the Reimerdes case). On his Web site, Corley noted that "there have been numerous reports of movie industry lawyers shutting down sites offering information about DeCSS. *2600* feels that any such suppression of information is a very dangerous precedent. That is why we feel it's necessary to preserve this information. . . . People with original copies of pages that have now been censored or removed are encouraged to send us copies for mirroring as well as links to additional information" ("DVD Encryption Cracked" 1999).

By the time the MPAA brought suit in New York, there was a concerted effort to distribute DeCSS on the Internet. Fritz Attaway, senior vice president for government relations and general counsel for the MPAA in Washington, DC, explained to the New York court that

When . . . the court [in California] declined to issue a temporary restraining order, members of the hacker community took this as a vindication of their actions. Displaying an "in your face" attitude, hackers taunted CCA and the MPAA by stepping up their efforts to distribute DeCSS to the widest possible worldwide audience. I am informed that one enterprising individual even announced a contest with prizes (copies of DVDs) for the greatest number of copies distributed, for the most elegant distribution method, and for the "lowest tech" method. (Universal v. Reimerdes et al., US District Court [SD NY 2000], Declaration of Fritz Attaway in Support of Plaintiffs)

Much like in the Bunner case, the EFF played a key role, providing legal defense for Shawn Reimerdes and the others. However, the EFF was much less successful with the Reimerdes case than it was with the Bunner case. Transcripts of the hearing for the petition to the court to enjoin Reimerdes show that the EFF lawyers had a very bad day in court. They failed to present supporting affidavits, had not really thought their arguments through, and failed to convince an openly antagonistic judge[1] that their defendants ought not to be enjoined. Unlike in the Bunner case, the EFF did not make cogent arguments regarding the issues of prior restraint and, it seems, could not navigate the DMCA in a fashion that convinced the judge that the defendants were acting within exemptions in the statute. Part of the problem was that the judge read the complaint quite narrowly and pressed the EFF to show how the defendants had in fact not violated the DMCA's anticircumvention and antidistribution provisions. Even though the EFF tried to argue that there may be some conflicts with fair use or even that some of the defendants might be protected by the DMCA's safe-harbor provisions, the court was interested only in knowing how exactly DeCSS was not a circumvention device outlawed by section 1201, which of course it plainly was.

Furthermore, the brazen statements made by many hackers distributing the DeCSS software—posting comments such as "Yes, you can trade DVD movie files over the Internet. . . . [T]he DVD Copy Control Association . . . are [*sic*] cocksuckers"—made a sympathetic interpretation of hacker motives almost impossible.[2] Posters of DeCSS technology came across as the work of nihilistic scofflaws, and without affidavits from computer scientists and other cryptographers the EFF could not give DeCSS any legitimacy. In contrast, the MPAA had significant evidence showing that the hacker and broader Internet community was engaged in an all-out campaign to distribute DeCSS as widely as possible. Comments from Fritz Attaway and Web pages calling for wide distribution made a damaging case against those distributing DeCSS.

DVD-Copy.com

share your DVD with the world

Yes, you can trade DVD movie files over the Internet, but be prepared, the favorite movies you might be interested in downloading could be a whooping four Gigabytes. A file size that even most connection speeds will have trouble completing in a reasonable amount of time.

You can break the encryption on any DVD and allow users to copy the contents of a DVD onto the a hard drive or alternative media!

DVD in the News

Notice: The DVD Copy Control Association are cocksuckers!

- *DVD Encryption Hacked (11-05-99)*
- *DVD Piracy : It can be done*
- *Why the DVD Hack was a Cinch*

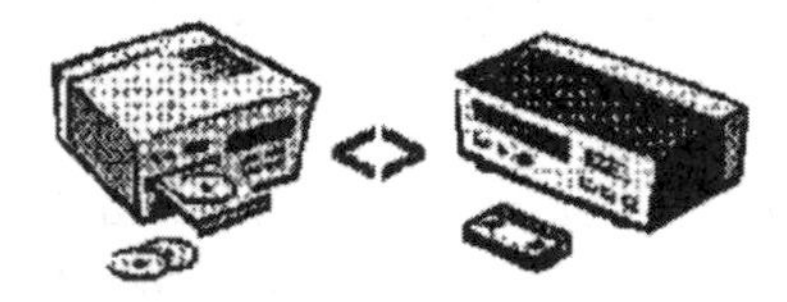

Figure 6.1

Screen capture of the home page of defendant Shawn Reimerdes's Web site dvd-copy.com. Note the text under its title. Image used with permission of site owner.

Even though some of the posts were indeed inflammatory (see figure 6.1), others had a more political bent. There was, in fact, a contest to see who could come up with the most creative way to distribute the DeCSS source code, and the distribution was couched in political, not nihilistic, terms. One contest flyer sarcastically read: "Winners of the contest will receive a copy of a DVD movie of their choice about an evil totalitarian society such as '1984' or 'Brazil' so they can watch the movie and thank God for their freedom," and another noted, "This is about the freedom of information, the right we all still have to learn how technology works—once this is gone there is no end to the kind of information that could be restricted because some conglomerate somewhere decides that its dissemination could cause them some grief" (quoted in Universal City Studios, Inc. v. Corley, US Court of Appeals [2nd Cir. 2001]).

The court was not interested in how hackers and users would define rights of access and use; as mentioned previously, the central question was whether DeCSS was indeed a circumvention technology in violation of the DMCA. With this question in mind, one cannot fault the court for pursuing the issue narrowly. The task of redefining those rights would fall on activists themselves as they set out to ignore the commands of Judge Lewis Kaplan and the Second Circuit Court of southern New York.

Defining the Issues

During the initial hearing on the case, the MPAA prevailed in court and got its preliminary injunction on Reimerdes and the others. During this hearing, the EFF tried to argue some of the more important issues pertaining to the digital rights movement. On the issue of code as speech, the EFF sought to point out the expressive aspects of code, in particular the part of the source code that contains programmer comments on the importance or rationale behind some lines of code. Second, the EFF noted that because the code was of interest to cryptographers, it must be available to be referenced and looked at by other members of the programming community. Enjoining the defendants from posting the code would interfere with this socially important action. The EFF argued that enjoining Reimerdes and the others from posting DeCSS before it had been determined to what extent DeCSS was expressive should be considered prior restraint. Therefore, the court was compelled to preserve the defendant's free speech and not issue the injunction.

Furthermore, the EFF claimed that DeCSS was protected under the exemptions in the DMCA for reverse engineering. Under these exemptions, an owner of a legally bought program can reverse engineer a copy-protection measure for the purposes of interoperability. The EFF told the court that DeCSS was such a program because it allowed for the design of DVD players for Linux.

The court rejected all of these assertions, viewing DeCSS as primarily a functional tool and not expressive in the way that analysis of prior restraint would require. The judge argued that the public good achieved by posting and distributing DeCSS was minimal. Even if there was some expressive element to the code, harm from its curtailment could not be compared to the harm done to the film industry, whose copyright was also a constitutionally important goal. Furthermore, the judge noted that curtailment of the code was not curtailment of the notes in the code, which could be posted.

As stated earlier, these arguments appear disingenuous. On the issue of reverse engineering, the judge read the statute quite narrowly. The statute explicitly notes that circumvention is for a program and that it has to be done for the sole purpose of interoperability. In its analysis, the court noted:

> First, defendants have offered no evidence to support this assertion.
>
> Second, even assuming that DeCSS runs under Linux, it concededly runs under Windows—a far more widely used operating system—as well. It therefore cannot reasonably be said that DeCSS was developed "for the sole purpose" of achieving interoperability between Linux and DVDs.
>
> Finally, and most important, the legislative history makes it abundantly clear that Section 1201(f) permits reverse engineering of copyrighted computer programs only and does not authorize circumvention of technological systems that control access to other copyrighted works, such as movies. In consequence, the reverse engineering exception does not apply. (Universal v. Reimerdes et al. [2000], Memorandum Order)

The statute appears inconsistent in its attitude toward reverse engineering of computer programs. What if, for example, the computer program is an anticircumvention measure? Does this imply that this program must not be reverse engineered? What if copy control is part of a larger program, and reverse engineering of the whole implies reverse engineering copy-protection devices?

The defendants' case was damaged early on by their inability to show evidence at the preliminary hearing that some of the activities in designing DeCSS were legitimately important to cryptography. They were also harmed by the court's perception that DeCSS was not designed in good faith. The judge wrote, "There is no evidence that any of them is engaged in encryption research, let alone good faith encryption research. It appears that DeCSS is being distributed in a manner specifically intended to facilitate copyright infringement. There is no evidence that [the] defendants have made any effort to provide the results of the DeCSS effort to the copyright owners" (Universal v. Reimerdes et al. [2000], Memorandum Order).

Expanding the Injunction and Continuing the Defense

Following the preliminary injunction, Reimerdes and Kazan dropped out of the case, choosing to discontinue distributing DeCSS. However, Corley stayed on and stepped up his efforts to distribute DeCSS on his 2600.com Web site, stating, "Help us fight the MPAA by leafleting and mirroring DeCSS" (as quoted in Reply Memorandum of Law in Further Support of Plaintiffs' Motion to Modify the January 20, 2000, Order of Preliminary Injunction, Universal v. Reimerdes, IS District Court [SD NY 2000]). Although he was enjoined from actively distributing the code, he linked to it heavily, providing more than four hundred links to the DeCSS application and source code on his own site, with some of those linked sites providing even more links (Universal v. Corley, US District Court [SD NY 2000], Supplemental Declaration of Robert W. Schumann). This act prompted the MPAA to request a modification to the injunction, asking the court to prohibit Corley not only from distributing the code, but also from linking to it.

In an initial declaration in opposition to expanding the injunction, Corley attempted to educate the court on the nature of hacker culture, hoping to the correct the pejorative meaning typically associated with the term. He noted:

> It is important to understand that the terms "hacker" and "hacking" as used by and about *2600* are not pejorative, but refer to the original sense of the term "hacker" as a person experienced or expert with computers and Internet navigation who is imbued with a spirit of imagination, innovation and exploration. In the traditional sense of the word, for example, "hackers" include professional security experts used by major corporations and governments to test the security of systems. (Universal v. Corley [2000], Declaration of Emmanuel Goldstein in Opposition to Plaintiffs' Motion to Modify the Preliminary Injunction)

Corley went on to point out how hacker practices have been misunderstood by society. He explained that the role of *2600* magazine was not only to instruct hackers, but also to "instill a sense of reality into the mainstream so that the actions of such people are judged in a more even-handed way and so that people aren't sent to prison for relatively minor offenses" (Universal v. Corley [2000], Declaration of Emmanuel Goldstein).

Corley's reason for posting DeCSS was entirely consistent with the hacker subcultural ethic and with many of the goals of the digital rights movement. Explaining why he chose to distribute DeCSS, Corley wrote, "When [DeCSS] was posted to the Internet, I recognized the importance of such a program to a variety of disciplines, including reverse engineering and open-source DVD player, cryptography, and in aid of legal consumer fair use. I was quick to show support for its existence and to condemn the attempts at forcibly quashing such knowledge" (Universal v. Corley [2000], Declaration of Emmanuel Goldstein). Finally, Corley complained bitterly about the court's inability to see legitimate alternative purposes for DeCSS and about the MPAA's bullying tactics toward other DeCSS distributors whom the court order had not enjoined. He noted:

> It is important to note that this entire issue is NOT about copying but rather about access. I believe it is entirely legal to use a DVD one has bought in a computer that one has bought. I oppose illegal copying but that has got nothing to do with DeCSS.

> . . . [The MPAA] has been sending cease-and-desist letters to some or all of the websites on our mirror list. The letters . . . are misleading and intimidating, since they suggest that the recipient "may" be subject to an injunction even though Plaintiffs know very well that the recipient is not. (Universal v. Corley [2000], Declaration of Emmanuel Goldstein).

An important development in this case was the EFF's recruitment of noted First Amendment lawyer Martin Garbus.[3] Garbus immediately filed to have the injunction vacated and began pressuring the MPAA to step up its discovery process. He made some important claims early on about the court's refusal to consider Corley's First Amendment rights properly. He noted, for example, that there was bias in the injunction against *2600* magazine because it was a hacker publication as opposed to a traditional news source. Garbus states that "had [the] plaintiff[s] sued *The San Jose Mercury News* or *The New York Times*, the resultant outcry would have been different" (Universal v. Corley [2000], Affidavit of Martin Garbus in Support of Defendant's Motion to Have Plaintiffs Post Security Bond). Richard Meislin, editor in chief of the *New York Times* digital edition, explained to the court that "the ability to refer readers to other Web sites that relate to a particular article is an integral part of the practice of journalism on the Web" (Universal v. Corley [2000], Declaration of Richard Meislin, Editor in Chief: *New York Times* Digital Edition).[4]

Garbus also argued that expanding the injunction to include hyperlinking would encounter some technological hurdles defined by Web browser technology. Linking involves annotating the text visible to the reader with some invisible HTML (hypertext markup language) commands; for example, one can type in the word *DeCSS* and have the word associated with an HTML code that calls for the specific Web address where DeCSS can be found. Or one can type out the Web address as part of the text and then tie the HTML to it. Some browsers are able to scan the text of a page looking for text that can double as an HTML command. Garbus noted that the browser can treat any Web address this way, even if not linked. Thus, even if hyperlinking were enjoined, the presence of the Web address in the text of an article would cause some browsers to treat it automatically as a live link. To prevent this, Garbus implied, the court would have to enjoin not only the conduct of linking, but also the very mention of any Web address that a browser may assume to be a live link (Universal v. Corley [2000], Brief Submitted by Media Defendant 2600 Enterprises).

Garbus also argued against the injunction on fair-use grounds and revisited many of the themes presented in the review of the formulation of the DMCA. As was done in the Sklyarov case, Garbus tied fair use to the First

Amendment. He explained that fair use strikes a balance against the monopoly on speech granted by copyright. He portrayed fair use as a right, an interpretation that had been heavily contested by copyright owners and the register of copyrights during formulation and review of the DMCA. Garbus wrote:

> CSS, which plaintiffs would have codified into law through its cramped reading of Section 1201, completely blocks access to the copyrighted material on a DVD, and prevents thereby any possibility that the right of fair use can be exercised with respect to that material. Congress did not anticipate or permit this . . . it is imperative to explode the favored analogy urged by plaintiffs, the MPAA and the DVD CCA, that no one has the right to break into a bookstore to make fair use of a book. In reality, the effect of CSS on fair use is to permit a publisher to prohibit a customer who has purchased one of its books from reading the work, except in a room constructed by a licensed builder, or under the lamp built by a licensed manufacturer. (Universal v. Corley [2000], Brief Submitted by Media Defendant 2600 Enterprises)

The notions of copy control and access control were central to this case. Deployment of the "breaking and entering" metaphor does not apply equally to both copy-control and access-control technologies. The exemptions that Congress made were for the first, not the second. Congress created an access right for copyright owners in the course of protecting access-control technologies. The digital rights movement argues that access rights, when under the protection of technological enforcement, preclude fair use.

Corley's Allies and Their Support of DeCSS

Besides the EFF's and Martin Garbus's involvement as defense counsel, Corley mustered an impressive array of support against the expansion of the preliminary injunction, with supporters coming from universities, law schools, and the computer industry (see table 6.1).

Many of Corley's supporters submitted statements to the court, and it became obvious how extensive resistance to the preliminary injunction had become. There were also close connections between the Bunner case in California and the Corley case in New York. Some defendants, for example, were named in both cases, and some of the people submitting statements of support for the defendants did so in both cases. Analysis of the court record in the Corley case shows an extensive network of activists coming together on the Internet to undermine the court's attempts at suppressing DeCSS. All of these activities were coordinated in an ad hoc fashion as participants took interest in the case and chose to "mirror"

Table 6.1
Corley's Supporters against Expansion of the Preliminary Injunction on DeCSS to Include Linking

Name of Supporter	Occupation/Affiliation
Pam Samuelson	Professor of law, Boalt Hall School of Law at the University of California at Berkeley
Richard J. Meislin	Editor in chief, *New York Times* digital edition
Charles R. Nesson	Professor of law, Berkman Center for Internet and Society at Harvard Law School
Chris DiBona	VA Linux Systems, Inc.
Bruce Fries	TeamCom New Media Consulting, LLC
John Gilmore	Founder, SunMicro Systems; EFF cofounder; founder, Cygnus Support/Affiliation with Red Hat Linux; cofounder, Cypherpunks, an informal educational and advocacy group devoted to advancement of privacy and security through greater knowledge and deployment of encryption
Lewis Kurlantzick	Professor of law, University of Connecticut Law School
Eben Moglen	Professor of law and legal history, Columbia University Law School
Matt Pavlovich	President, Media Driver, LLC, a consulting company that focuses on providing Linux video solutions to industry
Bruce Schneier	Chief technology officer, Counterpane Internet Security Inc., a cryptography consulting company, and author of one of the five encryption methods under consideration to become the US Advanced Encryption Standard
Barbara Simons	President, Association for Computing Machinery
Frank Stevenson	Computer research programmer, Funcom Oslo AS; first to publicly disclose cryptonalysis on the CSS ciphers
Dave Touretsky	Senior research scientist, Computer Science Department and Center for the Neural Basis of Cognition, Carnegie Mellon University
David Wagner	PhD candidate in computer science, University of California at Berkeley; cofounder, UC Berkeley's ISAAC Security Research Group
John Young	Operator, Cryptome.org Internet Library Archive

DeCSS or post the code themselves. The EFF played an important role in this regard by acting as a collection point of information and supplying defense counsel to those being sued by the movie industry. The defendants themselves were responsible for fomenting dissent against the injunction. The movement was coherently attacking the DMCA with both institutional and extrainstitutional tactics.[5]

Perhaps one instance above all others shows the determination to challenge the court's injunction: David Touretsky's Gallery of CSS

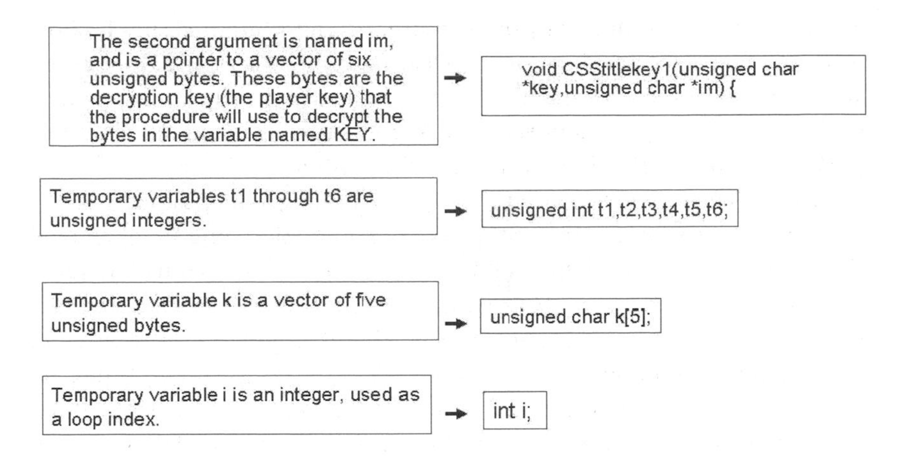

Figure 6.2
English to C translation of DeCSS. Adapted from David Touretsky's Gallery of CSS De-Scramblers (Touretsky n.d.).

De-Scramblers. Touretsky, a senior computer scientist at Carnegie Mellon's Center for the Neural Basis of Cognition, posted a collection of DeCSS variants to "point out the absurdity of Judge Kaplan's position that source code can be legally differentiated from other forms of written expression" (Universal v. Corley [2000], Deposition of Dr. David Touretsky).

On Touretsky's site, one could view DeCSS in various forms. For example, one could order T-shirts or ties with the DeCSS code inscribed on them, or one could view the DeCSS code translated into conversational English (see figure 6.2) with corresponding translations in the C programming language or in verse form (quoted in chapter 5).

Dr. Touretsky was very clear about the political nature of his gallery. In a written statement submitted to the court in support of Corley, he noted:

> I created the Gallery in response to the Preliminary Injunction issued by this Court. The Gallery consists of a set of files containing source code, textual descriptions of algorithms, and discussion of programs that can decrypt data that has been encrypted with CSS or that can recover the keys necessary for such decryption. . . . It is my belief that source code is expressive speech meriting the full protection of the First Amendment. This belief results in part from my experience as a computer science educator. . . . I am concerned that this Court issued an order prohibiting the defendants from posting source code for CSS decryption algorithms on the Internet. As a scientist, I feel it is imperative that anyone, not just academics, be allowed to participate in the ongoing analysis and improvement of encryption technologies. . . . My Gallery is a combination of scientific dialog and political statement. (Universal v. Corley [2000], Declaration of Dr. David Touretsky)

He also pointed out that many of the more than four hundred sites linking to DeCSS had a political bent. It was clear from his gallery and from other acts of support for Corley that the DeCSS case had inspired far more people than professional programmers. One high school graduate at the time posted DeCSS on his byline in his school's yearbook.

Besides subverting the court's order on linking and posting DeCSS, Touretsky made some incisive critiques of the injunction. Important among them was his assertion that it is impossible to have open discussion guaranteed by the First Amendment without being able to see and reference the DeCSS code. The court itself had said that its injunction was only on posting DeCSS and that the injunction was not meant to curtail its discussion. Touretsky pointed out the difficulty inherent in this distinction, telling the court, "My web site contains a copy of a textual description of the CSS decryption algorithm by the cryptographer Frank Stevenson. How is one to determine whether Stevenson's description is accurate? The only reliable way is to compare it with the source code for an implementation that is known to be correct because it has been compiled and run successfully" (Universal v. Corley [2000], Declaration of Dr. David Touretsky). The absurdity of an injunction on the "text" (the DeCSS source code) but not on discussion of the "text" was made clear during an exchange between Touretsky and the MPAA's lawyers during Touretsky's deposition.

MPAA: . . . How does Judge Kaplan's injunction affect the things that you express concern about in the sentences I just read?

Touretsky: Judge Kaplan has enjoined the publication of source code, and that will hinder the ability of people to discuss these algorithms. . . .

MPAA: Let's say you were interested in having a discussion with like-minded people about encryption technologies. You could do that through e-mail, couldn't you[?] . . . And you could, if you wanted to study source code in connection with the discussion of encryption technologies, you could send copies of the source code back and forth by e-mail, correct[?] . . . And, hypothetically, if like-minded people wished to discuss encryption technology and in the process study source code, at least hypothetically they could get together, form a private Web site and post the source code on that site, correct[?] . . . What is the difference between obtaining or studying the source code through that private Web site as opposed to obtaining it through the 2600 Web sites?

Touretsky: I think there are two differences. First of all, if discussion was [*sic*] restricted to this private Web site, people with a casual interest would not be able to obtain access to the material. And, secondly, if people were required to only discuss the source code on this private Web site, they would be denied their First Amendment rights.

MPAA: And how would that be, sir?

Touretsky: Because the First Amendment does not say that one can discuss things in private but not in public. (Universal v. Corley [2000], Deposition of Dr. David Touretsky)

Touretsky makes the point quite clear that when the court enjoined DeCSS but yet still claimed to protect speech, it was being inconsistent. A person with a casual interest in the topic would be completely excluded from discussing DeCSS, and those with significant interest would have to go to great lengths to be part of the discussion.

Themes in Statements by Corley's Supporters

As Corley's supporters submitted their declarations, it became clear that some themes consistent with the goals of the digital rights movement were continuously raised. First, they claimed that interpreting the DMCA's reverse engineering exemption[6] as narrowly as the court had done in Corley would have a chilling affect on innovation. David Wagner, a computer scientist at Berkeley, argued that without reverse engineering CSS, a DVD player for Linux would be difficult to build. He noted,

Reverse engineering is often tedious and time-consuming because computer programs are extremely verbose (by human standards), but it is not in principle difficult. . . . Based upon my experience and participation in and my observation of the academic and research communities at the University of California, Berkeley, reverse engineering is necessary, standard, and good for software and consumer electronic products containing encryption. . . . Reverse engineering CSS as part of this process was . . . a necessity: you cannot build a DVD player on which to play DVDs that are encrypted with CSS unless you know how CSS works and how to make the DVD play despite the presence of CSS. (Universal v. Corley [2000], Declaration of David Wagner)

Second, witnesses argued that there were other significant noninfringing uses for DeCSS, such as fair uses that had been shown to be permissible by the Supreme Court case Sony Corp. of America v. Universal City Studios, Inc. (464 US 417 [1984]). Noted legal scholar Pamela Samuelson, an early critic of the DMCA and a prominent figure in the network of activists and organizations that comprise the digital rights movement, pointed out that

the U.S. Supreme Court in *Sony Corp. of America, Inc. v. Universal City Studios, Inc.*, 464 U.S. 417 (1984) established the rule that copyright owners only have the right to control infringement-enabling technologies if they lack "substantial non-infringing uses." During the legislative struggle over the anti-circumvention provisions, Congress added a provision to the DMCA intended to preserve this standard by including section 1201(c)(2) in this law. Insofar as DeCSS has a substantial noninfringing use, such as the enablement of platform conversion, it should be permissible under both *Sony* and the DMCA. (Universal v. Corley [2000], Declaration of Pamela Samuelson in Opposition to Plaintiffs)

Charles Nesson of Harvard Law School pointed out that CSS interfered with fair use in the course of his teaching. He explained:

I frequently use multimedia in my teaching, adding audio and video to my classroom presentations to help tell the full story of a case or question. For example, in my Evidence class I use clips from "The Verdict" to illustrate closing argument, from "The Accused" . . . and from "My Cousin Vinny" to raise a variety of trial and ethical issues. Currently, I can assemble a series of selections from videotape to present at the time and in the order most effective for my lesson. I can store other segments on a computer for quick access if they become relevant to a discussion. If new works are made available only in DVD format, access controls such as those the studios seek to enforce here will prevent me from using such works as an effective teaching tool. I believe this is only one example among many fair uses that would be extinguished if [the] plaintiffs' reading of anti-circumvention were adopted. (Universal v. Corley [2000], Brief of Professor Charles R. Nesson as Amicus Curiae in Support of Defendants)

As noted in chapter 2, both the WGIP and the registrar of copyrights had addressed this view of fair use, noting that so long as users had the ability to access content in formats not protected by technological enforcement, then the copyright owners and the law were under no obligation to provide access to digital formats. However, some of Corley's supporters increasingly saw the ability to manipulate the kinds of uses of digital media described by Professor Nesson as a political issue. Bruce Fries of TeamCom LLC, a new-media publishing company, was motivated to write a book on the issue, to be titled "Fair Use: The Fight for Consumer Rights." In his statement to the court, he explained:

I conceived this book as a result of the recent court cases initiated by the Entertainment Industry in response to the Internet publication of the DeCSS source code. Fair Use focuses on the issues surrounding the fair use of copyrighted materials by consumers and researchers. It explains various forms of encryption and copy protection schemes that assume consumer dishonesty and prevent or restrict duplication of copyrighted works for legitimate purposes such as fair use and reverse engineering. . . . The book includes tutorials and source code for programs—including the source code for DeCSS—that enable consumers and researchers to circumvent copy protection schemes for fair use purposes. Obviously, it is crucial to the book's accuracy and credibility that I am able to publish the source code for DeCSS and other CSS descramblers. (Universal v. Corley [2000], Declaration of Bruce Fries)

Third, some of Corley's supporters complained that CSS was also being used to force consumers to watch unsolicited advertisements by enforcing a "no-fast-forwarding" section on DVDs. Computer scientist Matt Pavlovich noted that "[n]ot only does [the] CSS license prevent a player from fast-forwarding through those certain portions of a DVD that are marked 'no fast-forwarding,' including entire blocks of unsolicited advertisements, but [it] also . . . explicitly forbids its licensees from making a DVD player with that capability" (Universal v. Corley [2000], Declaration of Matt Pavlovich).

Last, according to some of the respondents on Corley's behalf, the injunction was in conflict not only with scientific norms, but also with hacker norms of sharing information and the norms of the free-software movement. Matt Pavlovich, who also was a defendant in the Bunner case, explained that "the reverse engineering undertaken to develop a Linux DVD player is also directly applicable (and necessary) to the development of DVD players for use on other open-source operating systems, such as NetBSD, OpenBSD, and FreeBSD. . . . The Linux DVD player will be 'open-source' and free of charge to consumers, as with most Linux open-source products" (Universal v. Corley [2000], Declaration of Matt Pavlovich).

Thus, themes of reverse engineering, fair use, scientific norms, innovation, and consumer rights (topics consistent with the digital rights movement) surfaced within the context the DeCSS case. This case became emblematic of what had gone wrong with digital copyright, and Corley's supporters were quick to point that out.

Perhaps no other claim made by the MPAA instilled greater pressure on the court than the claim that DeCSS ease of use would lead to rampant copying. The MPAA expended considerable resources to prove this point in hopes that the program's alarming simplicity would compel the court to perceive an acute threat to the film industry's copyrights. A deconstruction of these technical claims is helpful in illustrating how the narrative of fear regarding technological consequences was deployed to convince the court of some yet unforeseen copyright disaster.

Deconstructing the Fear of Rampant Copying

During court proceedings, some commentators expressed frustration with the claims that CSS was a serious protection mechanism meant to safeguard copyright. Like supporters of Andrew Bunner in the DVD CCA case, they felt that CSS was actually meant to control markets in regions. John Gilmore, cofounder of the EFF, explained:

> Many published DVD discs [*sic*] can only be decoded by a subset of DVD players. Under the name "region coding," the DVD industry has used its capability to create subsets to divide the world into seven regions and contracted to restrict the DVD players sold in each region to only play DVD discs intended to be sold in that region. The region coding system is not inherent in or necessitated by the design of the encryption system at issue, but is created by how the secret keys are administered. I believe that the DVD industry designed and implemented the region coding system in order to restrain global trade in DVD discs, so they can charge differential prices in different regions, and so that the release of particular movies can be delayed in particular markets, for the benefit of theater owners and the companies who rent them movies. (Universal v. Corley [2000], Declaration of John Gilmore)

Corley's supporters argued that if the movie industry were serious about protecting its content, it would have used a strong encryption scheme, not one such as CSS, which was known to be weak. Furthermore, they noted that digital security experts soundly rejected a tactic used in the encryption scheme—to hide the keys in the data of a DVD or security by obfuscation—because any committed individual could scour the data on the DVD and actually find those keys. Barbara Simons, a noted computer scientist and president of the Association for Computing Machinery, argued that "CSS uses only a 40-bit key, a length known to be breakable in a few minutes. It also employs a proprietary algorithm, rather than one that has been extensively tested in the public domain. The copy protection system relies heavily on obfuscation which, together with the carelessness of at least one licensee, appears to have created additional opportunities to break the system" (Universal v. Corley [2000], Declaration of Barbara Simons).

And Bruce Schneier, chief technology officer of Counterpane Internet Security Inc., noted that

> [t]he entertainment industry knew this was a problem, but failed to come up with a viable solution. Instead, DVD software manufacturers were supposed to disguise the decryption program, and possibly the playing program, using some sort of software obfuscation techniques. This is a technique that has never worked: there is simply no way to obfuscate software because it has to be on the computer somewhere, and is thus accessible to researchers, people engaged in reverse engineering, and the like. (Universal v. Corley [2000], Declaration of Bruce Schneier)

Given such technological failure, the defendants argued, claims that the entertainment industry had taken extensive measures to safeguard its content were hyperbole. Although this reasoning alone cannot be seen as a rationale for breaking technological mechanisms, it goes hand-in-hand with claims that testing and sharing information on security systems of this sort makes the security technology better and is consistent with scientific norms. In fact, every computer scientist who commented on the potential for expanding the injunction on Corley noted that such an action would curtail the free discourse of science. Andrew Appel, a computer scientist from Princeton, noted: "Based on my experience as a University professor and researcher, as a programmer, and as a serious participant on the internet [*sic*] since its birth, it is my opinion that scholarship and science, and the innovation that is so crucial to technological advancement and economic growth, will be seriously damaged by an interpretation of Section 1201 that would prohibit circumvention of security systems" (Universal v. Corley [2000], Declaration of Andrew Appel).

And, again, Bruce Schneier argued that "[DeCSS] is good research, illustrating how bad the encryption algorithm is and how poorly thought out the security model is, and must be available to cryptologists, programmers, and others as a research and intellectual tool through the normal channels—including, but not limited to, posting it on the internet [*sic*]. What is learned here can be applied to making future systems stronger" (Universal v. Corley [2000], Declaration of Bruce Schneier).

One important fact that came out during proceedings for expanding the injunction on Corley was that no cases of pirating commercial DVD movies by the use of DeCSS had been reported yet. Robin Gross of the EFF had met with Gregory Goeckner, MPAA vice president and deputy general counsel, at a panel discussion exploring litigation under the DCMA, "In the Trenches: Reports from the DMCA Battlefield," in New York. When asked if there was any evidence that DeCSS was being used for pirating, Mr. Goeckner had responded that "he was aware of no evidence of any actual piracy attributable to DeCSS" (Universal v. Corley [2000], Declaration of Robin Gross). Many speculated that there had been no pirating because DeCSS was actually too hard to use and because the technology was not yet available. Chris DiBona, a computer scientist for VA Linux, carried out an informal survey, asking members of his mailing list if they had been able to use DeCSS to copy a movie from a DVD. He reported:

> I posted general inquiries about DeCSS-related copying to the Linux, other open-source, and "hacker" (in the non-pejorative sense of individuals devoted to exploring the limits of the Internet) communities via a variety of mailing lists and websites, including but not limited to the SVLUG and DeCSS mailing lists and the opendvd .org website. These communities are made up of very skilled and technically capable people. None of the approximately 2000 people who responded to my e-mails and postings reported using DeCSS to make copies of DVDs. Indeed, only two people—both of whom insisted on strict anonymity as a condition of speaking with me because they feared reprisal from the MPAA—said that they were able to use DeCSS

to view DVDs they had purchased. However, both reported significant problems with playback. One experienced distorted video and both experienced stuttering sound. It's also worth noting that the individual who called the video "high quality" (although with bad sound) used a very expensive dual processing computer equipped with a great deal of random access memory. (Universal v. Corley [2000], Declaration of Chris DiBona)

Others showed that it would be uneconomical to pirate DVDs, calculating that,

[d]ue to the huge size of the files involved, making a verbatim copy of a DVD is impossible in essentially all easily transportable media commonly available today on personal computers. . . . Using the Internet to send or sell copies of stored movies is particularly unreasonable: uploading a single gigabyte over a 56K modem would take about 40 hours, so an entire DVD would take many days. The sheer bulk of the material makes it impractical for consumers to "pirate" DVDs using commonly available equipment. DVD-RAM has been out for a year, and its drives cost from $300–1000. But its discs only hold 2.6GB, cost $14 to $35, and are incompatible with everything else. . . . The available DVD-R recorder drives cost $3500–$5200. Blank recordable media for DVD-R are more expensive than buying pre-recorded DVDS. There is no incentive to copy a $15 DVD onto a $30–$60 blank DVD-R, rather than buying a second original at $15. (Universal v. Corley [2000], Declaration of John Gilmore)

Here it is important to point out an apparent inconsistency in the testimonies of many of Corley's allies, however. Computer scientists in particular made a point of noting that CSS was not good copy protection, that it was easy to break, and therefore posited that its primary purpose was to serve as the linchpin of a licensing mechanism that allowed for price control and regional distribution of content. But in statements such as those given previously, Corley's supporters said that copying was impractical and difficult because getting DeCSS to work required technical expertise and hardware with large storage capacity, which implies that CSS was at least creating a difficult enough barrier to copying. One might argue that these inconsistencies actually show that CSS was at least a "good enough" copy-protection system (good enough to stop the average consumer, who would have to incur costs to get DeCSS to work) as well as an effective price-control mechanism. But this argument conflates views that CSS is an objectively mediocre copy-protection scheme with arguments that it is practically effective. In the eyes of computer scientists noting that CSS was not sophisticated, the argument was necessary to show that the movie industry's claims to having made extraordinary efforts to protect content was exaggerated. Showing that it was practically effective was necessary to counter arguments that DeCSS would lead to rampant copying. In a sense, Corley's supporters made the argument that DeCSS, when thought of as a copying tool, was theoretically effective but practically not that useful. This gap between theoretical effectiveness and practical effectiveness opened the door for the central functional point of the case: that all DeCSS was really practically good for was to help Linux users develop DVD players for their computers. Critiques of the legal backlash against DeCSS from free-speech, fair-use, and other movement perspectives were important to frame this otherwise very technical issue (designing DVD players) within the movement's broader issues.

There is, however, something valuable to be learned from thinking of the theoretical and practical effectiveness of CSS and DeCSS. CSS is only practically effective when mapped onto a sociotechnical network that involves laws, economies of cost, and technological limitations. This relationship suggests that practical effectiveness is contingent and subject to the network of social and technical forces that define the possibilities of an artifact's use.

In Court: Themes in Testimony

The review of documents and arguments given so far has centered on the hearing meant to stave off the preliminary injunction on Corley. The MPAA's case against Corley went to court, and in this section I review those proceedings. Many of the themes highlighted during Corley's fight against the preliminary injunction resurfaced in the court proceedings. This is important because it shows that the movement's themes remained consistent and its arguments solidified as the case wore on. Corley's and Touretsky's testimony continue to be of particular importance. It defined issues for activists and shaped movement constituencies, which were expanding to include not only hackers and technology companies, but also the open-source community in general, the Linux development community in particular, and educators such as university computer science professors.

Corley, *2600*, and the Hacker Ethic

Eric Corley's testimony made clear how hacker attitudes toward information positioned DeCSS as technology symbolizing hacker beliefs. The hacker subculture is, as much as any subculture, constructed in opposition to dominant trends and mores within broader society. It is in principle subversive and constituted as a response to trends in society that attempt to establish legitimacy and ownership claims over information. Steven Levy first articulated the hacker ethic in his book *Hackers: Heroes of the Computer Revolution* (1984). He states that the hacker ethic is composed of six tenets: 1. "Access to computers—and anything that might teach you about the way the world works—should be unlimited and total. Always yield to the Hands-On Imperative," 2. "All information should be free," 3. "Mistrust Authority—Promote Decentralization," 4. "Hackers should be judged by their hacking, not bogus criteria such as degrees, age, race, or position," 5. "You can create art and beauty on a computer" and 6. "Computers can change your life for the better" (23–28). Because DeCSS was being suppressed, it became emblematic of the hacker ethic, and access to DeCSS became an end unto itself.

Corley, when asked why he founded *2600* magazine, noted:

> Well, I saw the need for information to be spread beyond computer nerds, people who are just simply calling into bulletin boards. . . . I thought it would be nice, because these people had so much to say, if there is a way, a forum for them to say it and actually read it on paper, and since nobody else was doing that, I figured why not. . . . We try and bring these people together, whether they are representatives of the government, people from different countries, 12-year-old kids who are just learning something, we try to bring them all together in the same room so they can share information and bring something out of that, and we find that many relationships are forged from this that last for a very long time. (Universal v. Corley [2000], Testimony of Eric Corley)

Consistent with other accounts of the hacker ethic, the information presented by *2600* has over the years had an element of playfulness with technology that illustrates an "I wonder if I can do this" attitude toward complex technological systems (see Raymond 2001 and Stallman 2002). For example, some articles have been instructive on how to explore or manipulate complex computer or communication systems. Titles of past articles in the magazine include "Snooping via MS Mail," "Cellular Interception Techniques," "Tips on Generating Fake ID," "AT&T's Gaping Hole," and "Cellular Network Detailed."

In his remarks, Corley showed a hacker's understanding of authority in information sources:

> What we mostly do [at *2600*] is print information. We have an editorial at the beginning of every issue where we expound on various thoughts, which is something that I write, and also in the replies to letters in the magazine, which is also printed every issue. If we give any kind of moral guidance or judgment, that's where it is, but in the actual articles themselves, it is more or less a compilation of material that is already out there, and we kind of present it to the people to show them this

is what people are saying, this is the information that's out there, this is how systems supposedly work or don't work. And people write in with corrections, they write in with additions, and we have a dialog going based on that. (Universal v. Corley [2000], Testimony of Eric Corley)

One of Corley's main points during the trial was that he was entitled to the same First Amendment protections allotted to journalists even though industry lawyers had suggested he was not. Importantly, Corley's view of journalism differed from the court's. A publication such as *2600* is rooted in a "see for yourself" practice that has become common in online reporting, where linking to original documents is part of the standard. Much like the hacker culture in which it is embedded, *2600* in its online version welcomes "crowd knowledge": the expertise of official contributors is parsed through the expertise of the contributing crowd. The authority is established when participants review claims and test or examine them. This is a marked shift in what would be considered typical journalistic approach to reporting facts and is in line with social computing practices that only recently have come into the mainstream in the forms of blogs and wikis.

The practice of letting a dialog establish the authority of knowledge claims runs against the practices and expectations of traditional journalistic reporting, where the tenets of journalistic methods often ensures that the "facts" of a story are related accurately. In fact, from developments in mass media, including scandals that have plagued prominent journalists such as Dan Rather in response to counterreporting by bloggers, such expectations of journalism have been shown to be unrealistic. Given ready access to primary material, readers and media consumers can make their own determinations of what is or is not fact or relevant to the accurate telling of a story. Therefore, the expectations of journalism in new media have shifted to where journalists are not only expected to present their version of the story, but also to "link" to the primary documentation that led them to their conclusions. This attitude is clearly informed by the hacker ethic, which calls for information transparency. Corley tried to illustrate this fact to the court and repeatedly noted that *2600* could not have presented a credible piece of journalism to its *particular* readership unless it also presented the DeCSS code.

The plaintiffs and the court never accepted this point, however, perceiving journalism in a traditional sense and not in the sense Corley was presenting it. The following exchange illustrates this difference:

Plaintiffs: You began posting DeCSS on your web site in November of 1999?
Corley: That's correct.

Plaintiffs: Is it your testimony that you did that as a journalist to write a story?
Corley: That's correct.
Plaintiffs: Could you have written the identical story without the posting, using the letters DeCSS as many times as you wanted in the story?
Corley: Not writing a story that would have been respected. . . . You have to show your evidence and in this particular case, we would be writing an evidence without showing what we were talking about and particularly in the magazine that I work for, people want to see specifically what it is that we are referring to, what bit of technology that doesn't work, what new advancement, what evidence do we have and simply saying that somebody else said something just won't cut it. So in this particular case, we pointed to the evidence itself which was already firmly established out there in the Internet world. We just put it up on our site so we could write our perspective on it and show the world what it was all about.
Plaintiffs: Getting back to the question, in November of 1999, was it possible for you to write a story about DeCSS on your web site, using the letters DeCSS next to each other as many times as you wanted, without posting DeCSS?
Corley: I will take another shot at it. I—basically, the story would not hold any value to our readers if we simply printed allegations without showing evidence. (Universal v. Corley [2000], Testimony of Eric Corley)

Corley's testimony showed a consistency with the hacker subculture that the plaintiffs chose not to accept and that the court dismissed in its decision, stating:

The name "2600" was derived from the fact that hackers in the 1960's found that the transmission of a 2600 hertz tone over a long distance trunk connection gained access to "operator mode" and allowed the user to explore aspects of the telephone system that were not otherwise accessible. Mr. Corley chose the name because he regarded it as a "mystical thing," commemorating something that he evidently admired. Not surprisingly, *2600: The Hacker Quarterly* has included articles on such topics as how to steal an Internet domain name, access other people's e-mail, intercept cellular phone calls, and break into the computer systems at Costco stores and Federal Express. (Universal v. Corley [2000], Opinion of Judge Lewis Kaplan)

Corley's testimony gave positive interpretations of hacker practices that the court saw as subversive to established regimes of authority and information ownership. Corley tried to frame his practices within accepted practices of journalism and freedom of information, but he apparently could not be understood by the establishment that he and his comrades were challenging.

Beyond proving consistent with the hacker ethic, Corley also saw posting and distributing DeCSS as a political act against censorship and called it an act of civil disobedience. He told his readers, "We have to face the possibility that we could be forced into submission. For that reason it's especially important that as many of you as possible, all throughout the world, take a stand and mirror these files" (Universal v. Corley [2000], quoted in Opinion of Judge Lewis Kaplan). The idea that posting code and linking it can constitute an act of civil disobedience tied directly into activists' belief that code is indeed a form of speech and that DeCSS in particular was political in what it stood for as well as what it did functionally.

Touretsky, the Hacker Ethic, and the Futility of Banning DeCSS

Corley's other supporters also framed their defense of DeCSS with the hacker ethic. David Touretsky, for example, painstakingly described his rationale for his Gallery of CSS De-Scramblers, noting that it was meant to convey two important points: (1) that the distinction between functional and pure speech in code is a fallacy; and (2) that code and its publication are an important way in which computer science professionals communicate with each other. His remarks were also consistent with the hacker ethic of free-information flows and an aversion to proprietary claims over it. Specifically, Touretsky reiterated the logical inconsistencies of separating code into functional speech versus pure speech. He stated that an injunction against some iterations of DeCSS would be meaningless in light of the fact that the code could be presented in a number of ways. In other words, if the court wanted to abridge distribution of the DeCSS source code, it would have to engage in the onerous task of abridging all the various forms in which it could be conveyed.

Touretsky used one interesting example to make the point that the DeCSS code had now taken many forms. Andrea Gnesutta had won an online contest for the most ingenious way of distributing DeCSS. Her winning submission was the image shown in figure 6.3. Gnesutta embedded the DeCSS source code in the image, and users were instructed to look at the source code of the image (distributed online) to extract the DeCSS code, which was embedded in the graphic file. In typical hacker fashion, a recursive system was used to distribute the image; the image itself said it was distributing DeCSS, and the code of that image *was* DeCSS.

Pavlovich and Why DeCSS Mattered to the Linux Community

Last, Matt Pavlovich's testimony was important in this case as well. He was a named defendant in the DVD CCA case and was the founder of LiViD, the open-source Linux Video project. His group made the most use of the DeCSS algorithm and authored some important components of the decryption schema. Pavlovich made clear in his testimony how important the DeCSS code was to the continued development of a Linux DVD player, a

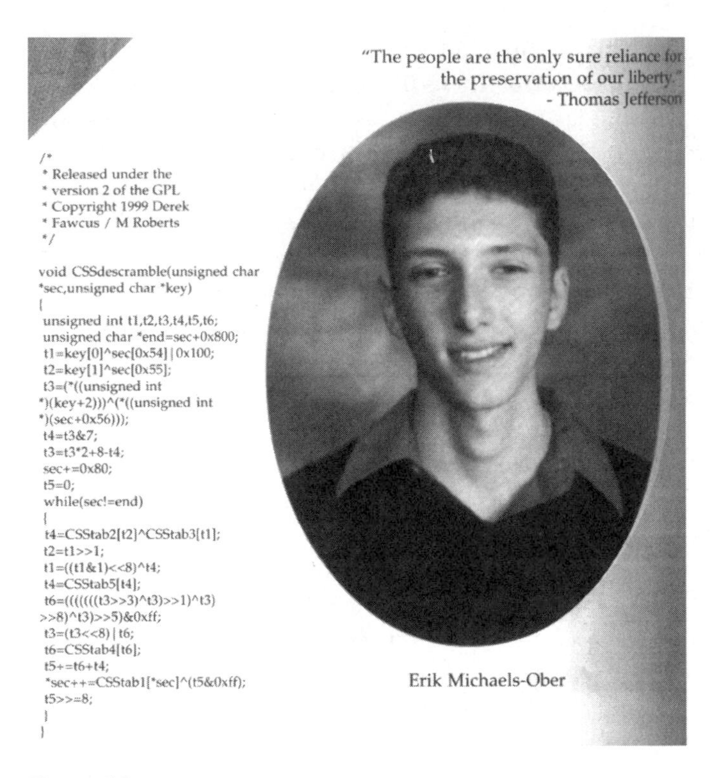

"The people are the only sure reliance for
the preservation of our liberty."
- Thomas Jefferson
/*
* Released under the
* version 2 of the GPL
* Copyright 1999 Derek
* Fawcus / M Roberts
*/
void CSSdescramble(unsigned char
*sec,unsigned char *key)
{
unsigned int t1,t2,t3,t4,t5,t6;
unsigned char *end=sec+0x800;
t1=key[0]^sec[0x54] | 0x100;
t2=key[1]^sec[0x55];
t3=(*((unsigned int
)(key+2)))^(((unsigned int
*)(sec+0x56)));
t4=t3&7;
t3=t3*2+8-t4;
sec+=0x80;
t5=0;
while(sec!=end)
{
t4=CSStab2[t2]^CSStab3[t1];
t2=t1>>1;
t1=((t1&1)<<8)^t4;
t4=CSStab5[t4];
t6=(((((((t3>>3)^t3)>>1)^t3)
>>8)^t3)>>5)&0xff;
t3=(t3<<8) | t6;
t6=CSStab4[t6];
t5+=t6+t4;
*sec++=CSStab1[*sec]^(t5&0xff);
t5>>=8;
}
}
Erik Michaels-Ober

Figure 6.3
Winning entry in the Great International DVD Source Code Distribution Contest. From http://www.dvd.zgp.org.

significant and legitimate use for DeCSS. Much of the resistance to the early injunction in the MPAA case and to the DVD CCA case came from the Linux and open-source communities. They saw these cases as direct attacks on their way of socializing and doing business. Open-source development was a foreign concept to many at the trial, however. The content industry representatives opined that open source would not amount to much of a business model, and they marginalized the importance of Linux DVD endeavors. Also, the court made much of the fact that because DeCSS had been designed for Windows, the Linux community could not claim it was intended to help in Linux DVD development. In response, Pavlovich testified that Linux machines had no way of reading the file structure on DVDs. DeCSS had to be designed for the Windows system so that Linux developers could use it to access DVDs and understand the encryption. They could then use this knowledge to develop the player for Linux machines. In spite of these reasons, the court and the plaintiffs complained that the cost of developing a Linux DVD application was too great. Windows was so widely available, the plaintiffs reasoned, that they could not risk having an application like DeCSS, which, even if it had legitimate uses, would give a great number of Windows users the ability to pirate DVDs. In this fashion, the Linux community's interests were marginalized and subsumed to the movie studios' interests.

The Corley Decision

It took only six days of trial for Judge Kaplan to find for the plaintiffs. In so doing, he issued a permanent injunction on Eric Corley with regard to the distribution and linking of DeCSS. As in the formulation of the DMCA, this decision was influenced not by actual data on personal pirating of movies or on data concerning damages, but rather on speculation of future losses.

Corley sought to appeal the injunction to the US Second Circuit Court and argued that the lower court had overrun his free-speech rights. Specifically, his defense noted

1. That the press can be enjoined from publishing truthful material because others, unrelated to the publisher, may someday use that material to violate the law.
2. That the press may be enjoined from even linking to such material.
3. That a lesser degree of First Amendment protection applies to expression that is "functional."
4. That Sect. 1201 [of the DMCA] trumps any right of fair use of digital works by preventing publication of technologies that allow fair users to have access to the works. (Universal City Studios, Inc. v. Corley, US Court of Appeals [2nd Cir. 2001])

Also, Corley's defense made note of the open animosity the court had shown the defendant during trial. Indeed, the court in all phases of the case had deployed some truly unfortunate metaphors to describe DeCSS and the work of hackers. It said that DeCSS was like a propagated outbreak epidemic where "individuals infected with a real disease are driven to seek medical attention, and are cured of the disease. Individuals infected with the 'disease' of the capability of circumventing measures controlling access to copyrighted works in digital form, however, do not suffer from having that ability. They cannot be relied upon to identify themselves to those seeking to control the 'disease.' And their self-interest will motivate some to misuse the capability, a misuse that, in practical terms, often will be untraceable" (Universal v. Corley [2000], Opinion of Judge Lewis Kaplan).

Comparing hackers to assassins, the court noted, "Computer code is expressive. To that extent, it is a matter of First Amendment concern. But computer code is not purely expressive any more than the assassination of a political figure is purely a political statement. Code causes computers to perform desired functions. Its expressive element no more immunizes its functional aspects from regulation than the expressive motives of an assassin immunize the assassin's action" (Universal v. Corley [2000], Opinion of Judge Lewis Kaplan).

These metaphors to describe hackers and their work served to alienate them further and paint their work as activities deplorable within society. But hackers and the open-source community had strong allies. On appeal, the case was argued by Kathleen Sullivan, the dean of Stanford Law School, and the briefs in support of Corley's position read like a *Who's Who* of legal scholarship and activism in the field of digital copyright (see table 6.2).

Despite the strong support from the academic and technical community, however, the appellate court affirmed the lower court's application of the DMCA, acknowledging that issues of equitable balance in copyright were to be settled by Congress. Corley and his supporters chose not to appeal the decision any further, noting that by the time any new decision was pronounced, DeCSS would be so widely available that an appeal would be irrelevant.

Table 6.2
Corley's Supporters in Universal v. Corley

Name of Supporter	Affiliation
Peter Jazsi	Professor, Washington College of Law, American University
Jessica Litman	Professor, Wayne State University
Pamela Samuelson	Professor, University of California at Berkeley
Ann Beeson	Associate Legal Director, American Civil Liberties Union Foundation
Christopher Hansen	National Staff Counsel, American Civil Liberties Union Foundation
Andrew Grosso	Association for Computing Machinery Committee on Law and Computing Technology
Julie E. Cohen	Professor, Georgetown University Law Center
Yochai Benkler	Professor, New York University School of Law
Lawrence Lessig	Professor, Stanford Law School
David A. Greene	Executive Director, First Amendment Project, Oakland, Cal.
Jane E. Kirtley	Professor, Silha Center for the Study of Media Ethics and Law, University of Minnesota
Erik F. Ugland	Graduate Assistant, Silha Center for the Study of Media Ethics and Law, University of Minnesota

Conclusion

In sum, this case put on trial the practices of academics such as David Touretsky, hackers such as Eric Corley and his readership, and Linux/open-source developers such as Matt Pavlovich. These groups proved difficult to suppress, primarily because they had the technological means to avoid the consequences of ignoring enjoinment and fomenting subversion. Furthermore, these groups had strong allies in institutions of higher learning, the media, and the legal profession. Even though Judge Kaplan sided with the plaintiffs, few of his stipulations would be met, and DeCSS would remain widely available.

But has DeCSS contributed to online pirating as feared? As noted earlier, economists and industry analysts wondered whether the projected losses in revenue due to DeCSS were accurately calculated, noting that individuals would not necessarily buy a movie if they could obtain it otherwise for free. Furthermore, the court relied heavily on presumptions of broadband penetration and the specter of Napster to inform its opinion. It noted that "while not everyone with Internet access now will find it convenient to send or receive DivX'd [compressed versions of movies] copies of pirated motion pictures over the Internet, the availability of high speed network connections in many businesses and institutions, and their growing availability in homes, make Internet and other network traffic in pirated copies a growing threat" (Universal v. Corley [2000], Opinion of Judge Lewis Kaplan).

Recent studies show that broadband penetration has steadily risen over the past few years. Yet Americans who share copyright-protected content have been increasingly doing so outside of the traditional peer-to-peer fashion—using iPods, for example (Rainie and Madden 2004). The content industry's efforts have contributed to curbing distribution of pirated content, yet this outcome may be due to the music industry's prosecution of individual users[7] rather than to suits against technology makers, who have not been handed complete defeats in the courts.[8] Also, enforcement may not be the only reason many users do not share files online. The availability of file-distribution businesses has helped funnel would-be consumers into legal channels, for example.

Illegal movie distribution on the Internet never reached the level of illegal online music sharing, for a variety of reasons. First, even though broadband penetration continues to increase, the time required to transfer a movie file, even when compressed, can be quite long. Second, peer-to-peer networks tend to be very unreliable in terms of the continuity of a given connection and in terms of the content actually available. Third, the compressed format must be viewed on a computer, which is an inconvenience for those who enjoy movies on their TV sets. Fourth, if a file is downloaded in a format that can be viewed on a television, such as Super Video Compact Disc (SVCD) or Video Compact Disc (VCD), it will be quite big (between 1.5 to 3.0 gigabytes); it must be burned onto multiple CDs, and many home DVD players will not play SVCD or VCD formats. Fifth, even when compressed, these videos take up about 650 megabytes of disk space. One of the reasons the music-distribution phenomenon took hold was that the mp3 format compressed music files to about 3.5 megabytes with little loss of quality. Therefore, distributors could have hundreds of songs stored on their hard drives without taking up much space. This is not the case for video, where a computer movie library would take up a great deal of space and be difficult for the average computer owner to maintain and share. Taken together, these technological realities, not the legal realities, have kept video from being shared in the fashion that the movie industry and the courts predicted. With the advent of mobile video (streaming), perhaps this state of affairs may change, but that remains to be seen. And last, even though there is illicit sharing of video online, there has yet to be convincing evidence that the movie industry has lost any revenue to it; its sales continue to be healthy. If such losses do occur, other emerging entertainment media that compete for the "entertainment dollar" must be accounted for.

Most recently, however, torrent technology (BitTorrent is discussed in chapter 8), together with effective video compression and higher bandwidth penetration, has overcome some of the technological hurdles that originally made video distribution impractical (primarily bandwidth limitations and the preservation of visual quality). Although it remains to be seen what the actual impact of such advances are, torrent technology signals yet another instance of technological systems that are made with explicit social-cultural intentions (sharing information efficiently on bandwidth-limited networks) and that run up against copyright owners' attempts to control distribution channels. In Sweden, this issue went to court when owners of the torrent site thePirateBay.org were found guilty of contributory infringement—not for distributing movies, but for distribution of torrent seeds that would allow users to distribute content (not just video) in a peer-to-peer fashion.

The court verdict not withstanding, users have framed the torrent system as a viable content-distribution system alternative to that governed by encryption, DRM systems, and centralized services such as Amazon.com and iTunes. The case study in the next chapter shows resistance by users and activists to such centralized systems of distribution and consumption. In iTunes and its DRM system, users and hackers found an important target for technological resistance, and they designed and used hacks with explicit values rooted in the movement's ultimate belief that users ought to have freedom in their consumption and use of digital content.

7 iTunes Hacks: Hacking as a Tactic in the Digital Rights Movement

A history of iTunes, the iTunes Music Store (iTMS), and the DRM system that once operated on music sold through Apple illustrates the development of hacks to iTunes and its access- and copy-protection measures. With this history, we can see a progression of technological resistance that began with DeCSS and AEPBR (consumer products that became politicized) and continued with hacks for iTunes (which were designed with both a political and technologically functional purpose).

The iTMS and the iTunes Client

On April 28, 2003, with much fanfare and publicity, Apple introduced iTunes 4.0 and quickly became the darling of the nascent digital media distribution business. The success of iTunes and the iTMS helped redefine how digital music, movies, and books would increasingly be distributed. The iTMS helped propel Apple's portable digital music device, the iPod, into the consumer consciousness—so much so that it currently dominates the personal digital music device market. iTunes 4.0 was designed to integrate with the iTMS and give the user access to catalogs for purchasing downloadable music.[1] However, despite the overall positive reception of the iTMS service, some users of the iTunes software opposed some of the technologically imposed conditions of the iTunes EULA and, more specifically, the terms of service agreement (TOSA) for iTMS.

The development of multiple technologies bent on undermining the iTunes EULA followed. Many of these technologies were designed with the explicit purpose of giving common users what designers felt were fair-use rights. These iTunes hacks are a marked departure from technologies such as the AEBPR and DeCSS in that whereas the latter technologies were designed with consumer needs in mind and with the still ambiguous understanding of the DMCA, iTunes hacks were designed in a way that was explicitly political: the designers created the hacks both to subvert copyright law and to provide functionality.

The iTunes EULA and Its Technological Enforcement

The iTunes EULA came to the consumer in the form of what has now come to be known as a "click-through agreement." At the time of installation, the user is shown a body of text in the software's installation window; she is asked to read it, and if she agrees with its terms, she is asked to click a button on the computer screen that will bring her to the next window, which will then complete the installation process. If the user declines to accept the conditions of the agreement, the program closes and is not installed on the computer. Because the iTunes software is capable of accessing the iTMS, the iTunes EULA makes some explicit demands about how the software ought to be used. The EULA states: "This software may be used to reproduce materials. It is licensed to you only for reproduction of non-copyrighted materials, materials in which you own the copyright, or materials you are authorized or legally permitted to reproduce. This software may also be used for remote access to music files for listening between computers. Remote access of copyrighted music is only provided for lawful personal use or as otherwise legally permitted" (2005).[2]

Having installed the iTunes software, a user can then access, via a link within iTunes, the iTMS, at which time the user is presented with another document, the iTMS TOSA. The user is once again asked to "click through" if she agrees to its conditions. The salient points of the TOSA include the following stipulations: "You shall be authorized to use the Products only for personal, noncommercial use. You shall be authorized to use the Products on five Apple-authorized devices at any time. You shall be entitled to export, burn or copy products solely for personal, noncommercial use. You shall be authorized to burn a playlist up to seven times" (2005).

Although not explicitly stated in the TOSA, there are two additional limitations on the use of songs bought on the iTMS that are of relevance. First, songs bought on the iTMS can be played only on the iTunes music player. Second, because the songs are encrypted with Apple's DRM system, they cannot be sampled, edited, or used in other creative fashions.

At the time of the iTunes 4.0 release in 2003, users criticized these limitations and the terms of the iTunes EULA as overly restrictive, unrealistic, and in violation of what they perceived as a fair-use right. The click-through TOSA with the iTMS stipulated that only three computers could be authorized to have copies of songs bought on the iTMS.[3]

Enforcing the EULA: FairPlay

Apple ensured that the terms of the iTunes EULA and the iTMS TOSA were adhered to by incorporating a method of technological enforcement. The iTunes software tracked the number of authorized copies of a song by collecting system-specific information about the computers in which iTunes songs were purchased and keeping that information stored in the iTMS servers. iTunes also encrypted the authorization code for the purchasing computer on the song itself. When a user tried to access an iTMS song on an unauthorized computer, the software asked the user if she wished to authorize the computer. The iTunes software would then connect to the iTMS servers to check for authorization. If the computer exceeded the number of authorized computers, iTunes would not play the song. In contrast, if the computer was cleared for authorization, the song was encrypted with the new computer's information, and a copy of that information was updated on the iTMS server (see figure 7.1).

The iTunes software would also keep track of the number of times a song was burned on a playlist to a CD. Because the encryption of the songs downloaded from iTunes was proprietary, the iTMS song would not play on any other software. It is interesting to note, however, that iTunes made no limitation on the number of copies and did not affect the playability of songs not bought on the iTMS. The system of checking

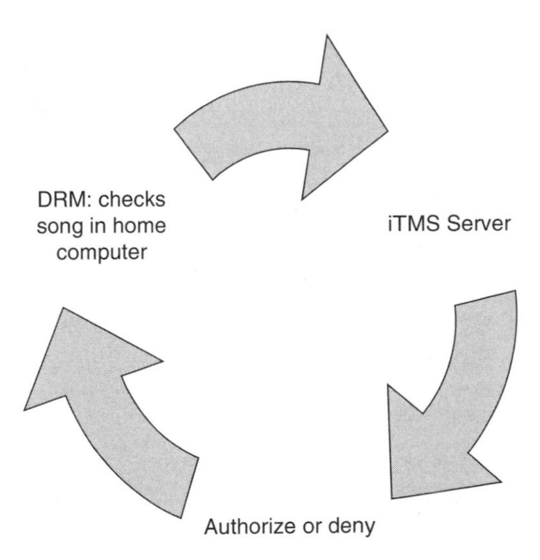

Figure 7.1
How Apple checks song authorization.

the number and authenticity of the iTMS songs was part of Apple's DRM system, FairPlay.

Early Circumventions: Dissention among the Faithful

The success iTunes enjoyed early on was a result of iTunes 4.0's connection with the iTMS and a series of new features introduced with that iteration of the Apple music client. Aside from having access to the catalogs of the iTMS, iTunes 4.0 introduced the file-streaming feature Rendezvous, which allowed one user in a household to stream content to any other Macintosh on the home network. Content could be accessed via the streaming feature so long as the user provided the iTMS user name and the password used to buy the song on the iTMS (Borland 2003a; P. Cohen 2003). For example, if a Mac in one part of the house had music on the hard drive, and a user wanted to listen to it on a Mac in another part of the house, the Rendezvous feature would stream the music between the two machines. However, the music was not "downloaded"; it was simply streamed between machines, much like a Webcast via Internet radio.[4] It was this simple streaming feature that users first appropriated for purposes other than what Apple had intended.

About two weeks after the release of iTunes 4.0, users figured out that they could open their home networks to users on the Internet. Various Web-based groups soon started providing search services and applications that allowed iTunes users who provided their ISP address to share their music libraries with other iTunes users via Rendezvous, thereby essentially turning iTunes into an "on-demand" Webcast application. These services and applications included: (1) a music-sharing database that held the ISP addresses of users willing to stream their music libraries via the Internet (Heidi 2003); (2) ServerStore, a downloadable database and application that listed ISP addresses of users sharing music and gave the user the option of publishing her own address (Borland 2003b; "iTunes 4 Tip" 2003); and (3) iTunesTracker, an application that would allow the user to connect to a central server and browse other users' shared libraries, thus functioning in the same fashion as the early Napster, but having only streaming capabilities (Borland 2003b; "iTunes Tracker" 2003b).[5]

It had taken Apple considerable work to negotiate licensing of music catalogs from record labels, and the linchpin of the agreement was the assurance by Apple that it could preserve the access rights delineated by copyright owners. Webcast licensing is required by law, even for amateur Webcasters, and capturing the music stream as a download was in 2003 and continues to be a possible means of creating copies of content. The new uses that iTMS customers were exploring thus violated the promise that Apple had made to music companies. The EULA and the TOSA in effect delineated how music companies wanted their music distributed and used. The key here is that those guidelines were not congruent with what users had come to expect in music they perceived as "bought." The EULA and the TOSA essentially rented the music to consumers, a point that did not come clearly through in the extensive language of those documents. The prospect of Apple's running afoul of the delicate agreement it had with record labels suddenly became palpable, and Apple moved quickly to address the appropriation of Rendezvous. Unfortunately for Apple, hacks that turned iTunes into a file-sharing program, a type of program that goes beyond mere streaming, had been developed even before the release of iTunes 4.0.

First among these hacks was iCommune, designed by software developer James Speth months prior to the release of iTunes 4.0. iCommune was designed to allow users on a network to stream *and/or* download music to other Macs on the same network, unlike the Rendezvous hack, which only streamed the content. iCommune integrated with iTunes as a plug-in and used some of iTunes' proprietary source code. As a result, Speth could not distribute the code widely without legal ramifications. But on April 14, 2003, just two weeks before the release of iTunes 4.0, Speth released his project as an open source on SourceForge.org, having reworked it to contain none of Apple's proprietary code. With iTunes 4.0's release in mind, Speth noted in an interview with the online press that his next step would be to integrate iCommune with the Rendezvous feature. Such a feature, in conjunction with the various ISP address databases available at the time, would have turned iTunes into a peer-to-peer file-distribution application, where users could find each other over the Internet and share content with one another (Fried 2003b). The iCommune site stated: "iCommune is a standalone open source application for Mac OS X that extends Apple's iTunes to share your music over a network. You can share the music in your iTunes library and access other iCommune music collections. iCommune music collections appear as playlists in your iTunes window. You can browse through them, and choose to stream or download the music they contain" (Speth 2005).

Versions of this software are currently still available from SourceForge .org, the open-source clearinghouse site, and its authors claim compatibility with iTunes 3.0 or higher (Borland 2003b; Speth 2005).

Speth, however, was not the only hacker who wanted to expand the Rendezvous feature. A series of other programs quickly followed the launch of iTunes 4.0. On May 11, David White released iLeech; on May 12 Gus Holcomb released iSlurp; and on May 17 Tobias Lernvall released iTunesDL (Heidi 2003; Holcomb 2003; "iTunesDL" 2003a; White 2003). All were applications that would allow a user to exploit the Rendezvous architecture to download music from other users.

Within two weeks of iTunes 4.0's release, there were at least eight services and applications that exploited the Rendezvous feature and expanded it beyond what had been intended by Apple developers.[6] Later applications allowed users with Mac or Linux machines to stream music from one to the other. Hacks that would allow this type of interoperability between operating systems included TunesBrowser and AppleRecords, released by David Hammerton in March 2004 and Chris Davies in May 2004, respectively (they were released nearly a year after the release of iTunes 4.0).

In 2003, however, Apple had to respond quickly to the growing possibility that iTunes would be appropriated for file sharing. Within a month of releasing iTunes 4.0, it released iTunes 4.01, plugging the hole that allowed users to stream and download music over the Internet (Fried 2003a). iCommune then responded to the iTunes 4.01 update with iCommune401(ok) which reestablished Netwide streaming in iTunes. Other less technical users did not develop hacks but instead devised "work-arounds" to iTunes 4.01's limited-sharing features. One user suggested simply having a dual installation of the iTunes software:[7] when a user wanted Internet streaming, she simply had to run iTunes 4.0 instead of 4.01 (Khaney 2003).

Even though some of these early hacks and work-arounds to the iTunes software allowed copying and streaming, the iTunes 4.0 DRM system remained secure. In its press release at the time of the iTunes 4.01 upgrade on May 27, 2003, Apple noted:

> The new iTunes 4.01 update limits Rendezvous music sharing to work only between computers on a local network (its intended use) and disables music sharing over the Internet. The iTunes . . . Music Store has been very successful to date, and the mechanisms we put in place to secure that music against theft are working well. . . . Music purchased from the iTunes Music Store can only be played on up to three authorized Macintosh computers, and there has been no breach of this security. (quoted in Fried 2003a)

Circumventing the iTunes DRM system would take another six months, and it would signal an escalation of the nascent war between Apple and some of its users. The iTMS and iTunes expanded to the Windows operating system, and so did the hacks that sought to circumvent limits on use. It is worth noting that Rendezvous remains part of the iTunes software today and seamlessly integrates a user's music library into the available playlists on other computers running iTunes with the users' home network. Therefore, downloading is still quite possible within these networks, although broadcasting to the Internet is not. The iTunes sharing option remains a concern for content owners because larger networks in universities make this type of sharing easy. With the release of iTunes for Windows, these concerns became more acute.

iTunes for Windows and QTFairUse: The First Volley

On October 16, 2003, Apple released iTunes for Windows, bringing the PC market within reach of its rapidly growing digital music distribution business. Less than two weeks later, on October 26, twenty-year-old college student Bill Zeller released MyTunes. MyTunes did for the Windows version of iTunes what iSlurp, iLeech, and iCommune did for the Mac version. Working within the limitations of the new version of Rendezvous, Zeller's software made it possible to download music within the same network and save that captured stream in the popular mp3 music format.

In spite of this circumvention, the DRM system on iTunes songs bought via the iTMS remained intact on Windows machines, and MyTunes could only generate mp3s for unprotected content, such as songs ripped from commercially available music CDs (Borland 2003c; Dalrymple 2003; Menta 2003). In an interview with the online press, Zeller acknowledged the potential legal tangle that such software could create:

> I would like to think they would go after those infringing copyrights and not those abiding by them. . . . However, although MyTunes can be used for legal ends, I understand how they (Apple and/or the RIAA) might have a problem with the software. I would like to think the responsibility to act in accordance with the law is on the user. Authors of software should not have to baby-sit the user in order to insure legal compliance. (quoted in Dalrymple 2003)[8]

Within a matter of months, Zeller's program had been downloaded more than thirty thousand times from the popular download site Download.com. Zeller himself reported that it had been downloaded one million times from other sites "mirroring" the program. However, by March 2004, MyTunes had vanished from the Internet due to a hardware failure on Zeller's computer. The failure erased all the source code and prevented further critical updates to the software to counter Apple's newer versions of iTunes that blocked the functionality of MyTunes (Borland 2004b; Menta 2003). On the Mac, GetTunes, developed by programmer Gregor Triplehorn as an open-source project, was released in May 2004 (Triplehorn 2004a, 2004b). It was billed as "[a] Mac version of MyTunes, a small application that allows users to download music from local Rendezvous-shared music libraries (instead of streaming the songs)" (Triplehorn 2004a).

Five months after MyTunes vanished from the Internet, David Blackman, a Stanford University programmer, released OurTunes, a variant of MyTunes for Windows with the same functionality.[9] The OurTunes project, hosted on SourceForge.net, is as of this writing still available for both the Mac and the Windows versions of iTunes. Like MyTunes, OurTunes made it possible to download songs from other members on a local area network sharing songs.

The hacks that plagued Apple iTunes for Mac and Windows shortly after the introduction of the iTMS did two things: it opened up the streaming feature, Rendezvous, to computers outside home networks; and it added a downloading feature to iTunes. These hacks engendered a sort of "code war" between Apple and hackers, with new versions of iTunes being released to counter hacks and new hacks being produced in response. Yet no breach of the Apple DRM system was made in the battle until hacker Jon Johansen did so on November 17, 2003, just a month after the release of iTunes for Windows.

Jon Johansen was not new to controversies surrounding access-control and copy-control technologies. In 1999, working with the hacker group MoRe, he had released DeCSS, a tool that incorporated the decryption algorithm for movies on DVDs, making it possible to play them in players not authorized by the MPAA and the DVD CCA.[10] On November 17, 2003, Johansen released the program QuickTime for Windows AAC Memory Dumper, or QTFairUse, as it came to be known. QTFairUse exploited a weakness in the algorithm for encrypting protected Advanced Audio Coding (AAC, a type of high-quality audio file) files bought via the iTMS. Songs bought from the iTMS have the ".mp4" file extension. Mp4 files are files that have been compressed using the AAC algorithm, have the MPEG Layer 4 audio encoding, and are protected by the iTMS DRM system FairPlay. It is possible to have AAC compressed files that are not protected by the iTMS DRM system but that are still encoded using the MPEG Layer 4 encoding, in which case these files would show up as ma4 files when viewed, for example, in the Windows file explorer. QTFairUse captures the protected AAC (mp4) file after it is decrypted but before it is processed to be played as audio, and it "dumps" a raw data file on the hard drive of the computer playing the song. That raw data file is not protected by Apple's DRM system because it was derived from the AAC file after it had been decrypted (figure 7.2).

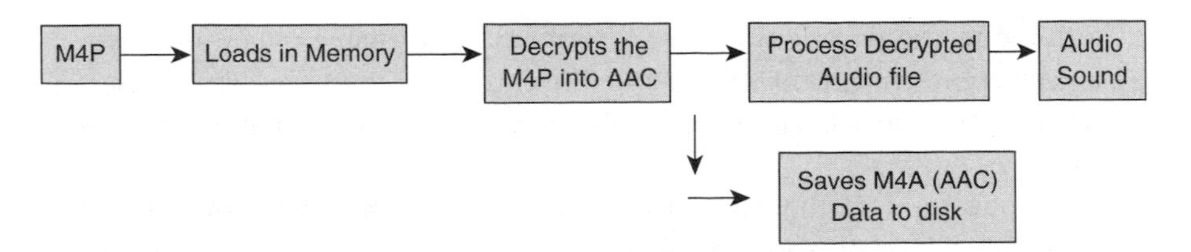

Figure 7.2
Algorithm for the operation of QTFairUse.

Shortly after the release of QTFairUse, Mac hobbyists on the Net had extensive discussions on the method by which QTFairUse achieved its ends. It is worthwhile to note that capturing encrypted audio has been an ongoing practice carried out before the advent of QTFairUse. The easiest way to capture the encrypted audio was well known among users of the iTunes software and was suggested by a contributor to *MacRumor* forums. The method requires simply burning a CD of the iTunes-protected media and then ripping (copying) it back onto the hard drive in mp3 format (or whatever other format the user desires so long as the format conversion software is available). Another method to capture the encrypted audio file was to download any number of audio-capture programs that would record the music as it was played by the sound card. Both of these methods unfortunately resulted in a loss of quality because they involved multiple conversions of the digital media format, from protected AAC (iTunes), to Audio Interchange File Format (AIFF) to unprotected AAC or mp3 in the case of copying to CD and back. In the case of capturing the data as the sound card plays it, the digital media file is converted from protected digital format to unprotected analog and then to unprotected digital. As figure 7.2 illustrates, the benefit of QTFairUse was that it captured the AAC data after they had been decrypted by iTunes but before they were processed to be played on the iTunes player. The quality of the sound from these captured data was therefore as good as that of the encrypted file (Chaosmint 2003; Geekpatrol 2003; "QTFairUse?" 2003).

Using QTFairUse was not without its complications; at the time of its release, many of the more popular music-player applications (Windows Media Player and Winamp, for example) did not support the AAC format. Furthermore, the raw AAC format was unreadable in the fashion "dumped" by QTFairUse and needed to be encoded with MPEG Layer 4 encoding, which would then generate an ma4 file. For all practical purposes, it appeared that QTFairUse users were stuck with a decrypted file that no player could play. Some users were able to generate the proper formatting and encode the raw AAC data into a readable m4a format, yet for most users QTFairUse remained technically complex. These users simply chose to use other methods of relieving their songs of Apple's DRM system.

Despite the fact that only a minority of computer users were skilled enough to decrypt and encode the raw AAC data, the importance of QTFairUse cannot be denied. It was the first program to render protected iTunes files into an unprotected format and, as such, was a crucial "proof of concept" that the iTunes DRM system, FairPlay, could be circumvented while maintaining the sound quality of the music.

All these machinations on the part of users as well as the development of QTFairUse required that the song be played by iTunes. Many of the users commenting on message boards and technology forums expressed the desire to want to play legally bought songs on devices other than iTunes and to want to play them in more than the three computers that the iTunes DRM system initially authorized. Whether those devices were Linux machines or other media players, many users felt that restricting the playability of legally bought music to one piece of software or to some limited number of personal devices was not what they bargained for when they bought a song from iTunes. One user quoted in the online press noted:

> Applied to a physical media . . . the idea of the DRM is this: You can play the CD on three designated CD players that support the DRM. . . . [I]t will play ONLY on xyz brand cd player and only three of those that you pick. Yes, you have to stick to that brand of cd player (the iTunes player, the supported OS of iTunes, no unix support in sight) and too bad if you have a fourth one in the bedroom. It's not gonna play in your second car's player either. Nor in the kitchen. Nor on your neighbor's player. Nor can you trade it on the used market when you're tired of listening to it. (Orlowski 2003, emphasis in original)

This user's statement is exemplary of the various frustrations that some consumers faced. It also illustrates the blurred boundaries between what users recognize as fair use and what the law delineates as a legal use of media. It illustrates the difficulties in implementing the "access right"[11] in digital intellectual-property distribution, and it goes to the heart of the digital copyright debate by pointing to the deviation between what some users see as a normal and legitimate practice and the practices that the law deems acceptable. The reason why users chose to use QTFairUse and other circumvention technology was that it allowed them interoperability and expanded their rights of access.

The VideoLan Client and drms.c: Escalation, Part 1

In January 2004, Johansen and fellow developer Sam Hocevar struck again: they reverse engineered the FairPlay DRM system for Windows. Their hack was modestly titled "drms.c" and was made available as iTunes playback support in the unofficial development library of the open-source media player VideoLan Client (VLC).[12] The software had two important uses: first, it allowed users to play iTMS songs on a player other than iTunes; and second, it allowed users to play songs bought from the iTMS on any number of machines, regardless of what the iTMS TOSA said. Thus, after just eight months of operation, FairPlay encryption was obsolete for those in the technological know.

On March 2, the VLC team released VLC version 0.7.1, which integrated Johansen and Hocevar's work into the project's official release. As reported by Andrew Orlowski for *The Register*, Johansen deduced that the key for encrypted songs was "derived from four factors: the serial number of the 'C' drive on the host computer [the computer containing the iTMS song], the host system BIOS,[13] the CPU [central processing unit] name for the host computer and its Windows Product ID" (2004). A user who wanted to play a FairPlay-protected song on an unauthorized machine first had to play it using VLC on a machine that was authorized. After the song was played, VLC would write the song key to a file on the hard drive (see figure 7.3). The user could subsequently listen to the song on unauthorized machines running VLC by simply copying the song and the key to the new machine (figure 7.4).[14] The technologically enforced limit on the number of computers that could play the song was thus defeated.

Importantly, drms.c remained tied to the VLC, and songs had to be played by VLC before keys could be generated. However, just a few weeks after drms.c's release, another programmer uncoupled it from the VLC and thus expanded the cross-platform playability of songs bought on the iTMS.

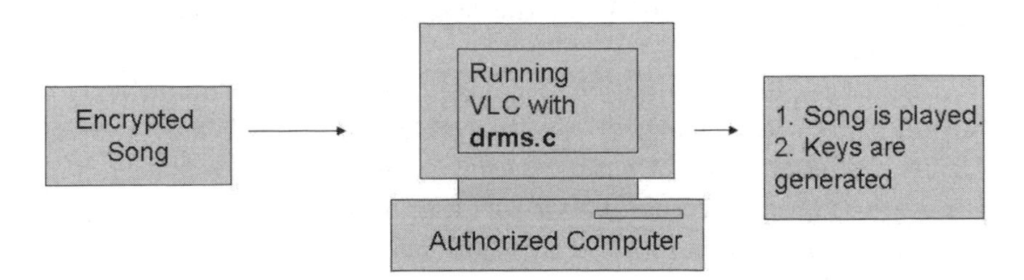

Figure 7.3
How drms.c plays and produces a song key.

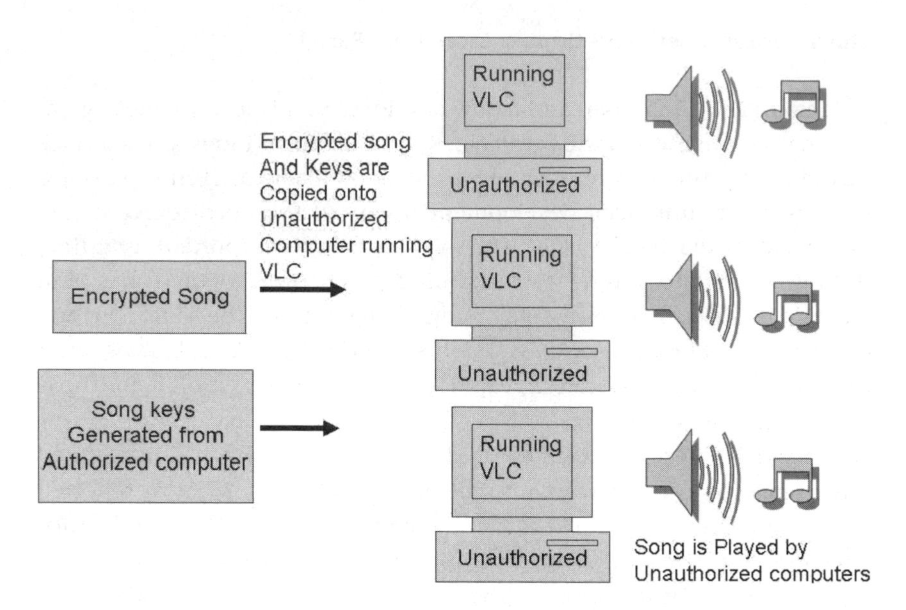

Figure 7.4
How drms.c incorporates into VLC functionality, allowing multiple machines to play a song.

PlayFair: Escalation, Part 2

PlayFair (not to be confused with FairPlay, the DRM for iTunes), developed by an anonymous programmer as an open-source project and released April 6, 2004, used Johansen and Hocevar's reverse engineering work but was a stand-alone application, as opposed to being incorporated into VLC. According to the developer, the application functioned on most UNIX-type systems (presumably Linux), Mac, and Windows systems. PlayFair was initially a command-line interface application for both Mac OS X and Window machines (see figure 7.5). Later iterations of PlayFair included a graphical user interface (GUI) for the Mac OS X and Windows (Hymn Project 2004b).

Importantly, PlayFair not only decoupled song playback from the VCL but also produced an unprotected, playable AAC output file of equal quality to the protected AAC file.[15] PlayFair improved on Johansen and Hocevar's VLC hack because it allowed a user to play a decrypted song on any number of machines *without having to copy the song keys along with the song* (figure 7.6).

The FairPlay system on the Mac OS X was not as easily defeated, but PlayFair did give Mac users some options. At that time, users who wanted

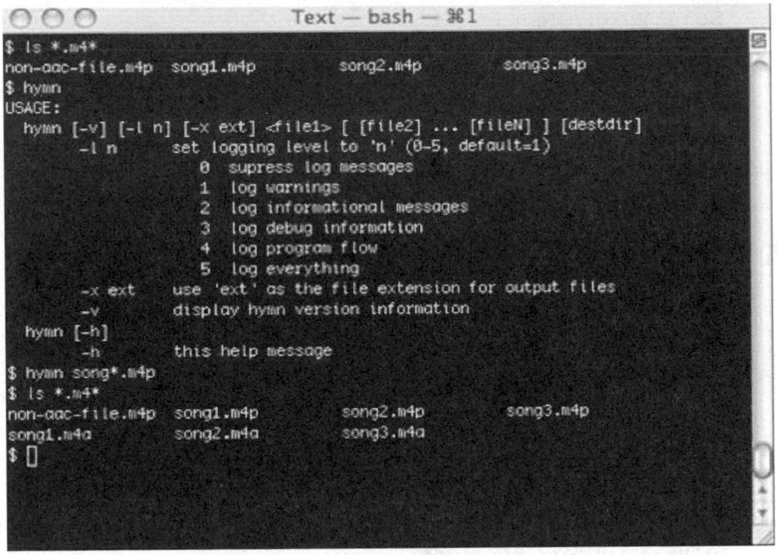

```
Text — bash — ⌘1
$ ls *.m4*
non-aac-file.m4p  song1.m4p         song2.m4p         song3.m4p
$ hymn
USAGE:
  hymn [-v] [-l n] [-x ext] <file1> [ [file2] ... [fileN] ] [destdir]
        -l n      set logging level to 'n' (0-5, default=1)
                       0  supress log messages
                       1  log warnings
                       2  log informational messages
                       3  log debug information
                       4  log program flow
                       5  log everything
        -x ext    use 'ext' as the file extension for output files
        -v        display hymn version information
  hymn [-h]
        -h        this help message
$ hymn song*.m4p
$ ls *.m4*
non-aac-file.m4p  song1.m4p         song2.m4p         song3.m4p
song1.m4a         song2.m4a         song3.m4a
$ 
```

Figure 7.5
Early interface for the PlayFair program.

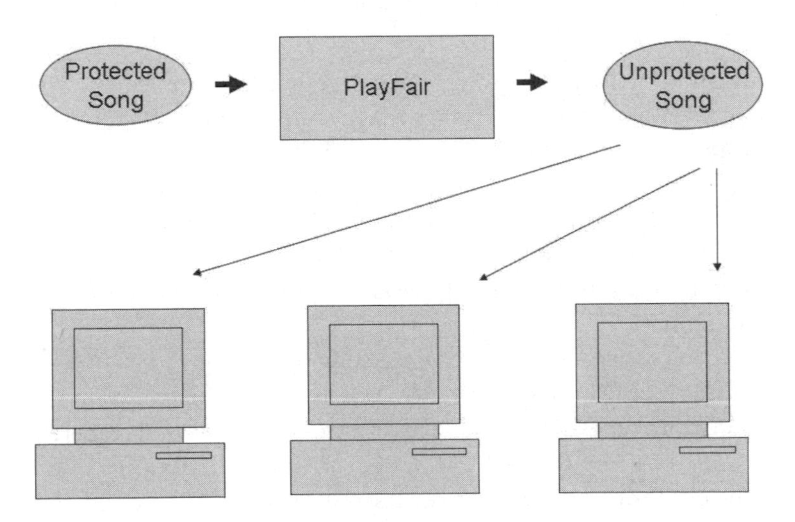

Figure 7.6
PlayFair produces unprotected songs and allows them to be played on any computer with any player.

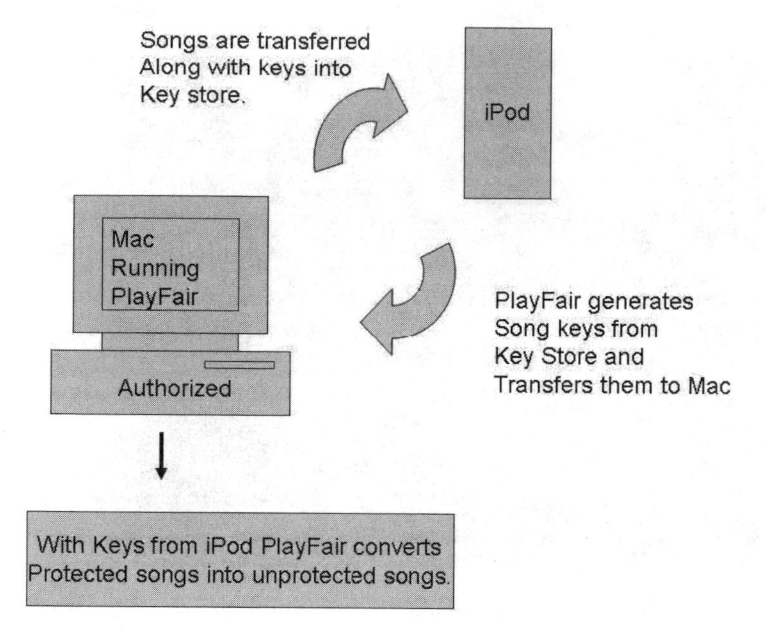

Figure 7.7
PlayFair uses iPod to generate song keys for Mac users.

to strip the DRM system from songs on a Mac needed to have an iPod mounted on the Mac with all the songs they wanted to decrypt stored on the iPod from an authorized computer. Because the iPod needed the song keys to play the stored music, iTunes loaded those keys onto a hidden folder on the iPod known as the "key store." PlayFair searched the iPod connected to the Mac, accessed the "key store" folder, and used those keys in conjunction with Johansen and Hocevar's software to strip the songs on the Mac (figure 7.7).

The GUI made conversion of a protected song somewhat easier than for the early Windows DOS interface. A user could simply drag and drop the protected song into the PlayFair interface, and the program would generate a new file that was free of the DRM system (figure 7.8) (Hymn Project 2004b).[16] The user, after having generated the unprotected AAC file encoded in MPEG Layer 4, could then play it on players that supported the format. Windows users, although not having access to an early GUI, had the advantage of not needing to plug in an iPod.

During the two years following the release of iTunes 4.0, the process of stripping songs of their iTunes DRM system became increasingly simple. For example, at the time of the QTFairUse release, users were having some

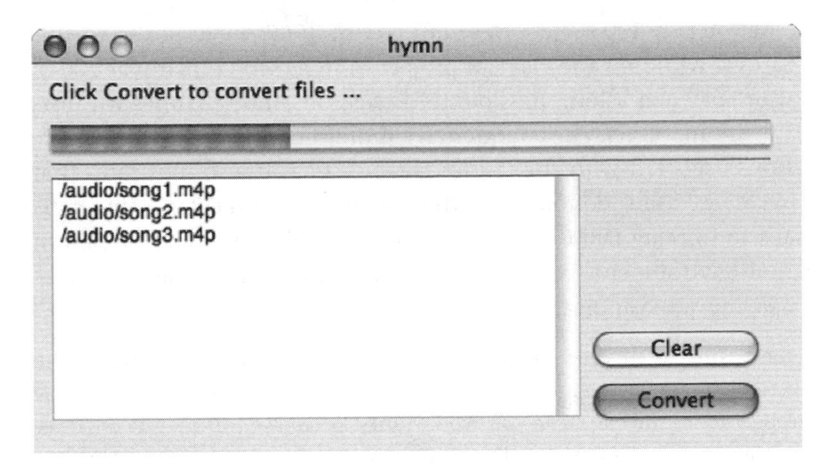

Figure 7.8
PlayFair GUI after its name had changed to "Hymn." Image taken from http://www.hymn-project.org.

difficulty playing the raw AAC file. By the time PlayFair was released, the media players Winamp, Free App a Day (FAAD), and VLC were capable of easily playing these songs. Thus, it was finally possible to convert a protected AAC file to an easily playable unprotected AAC file, doing away with the potential loss in quality due to conversions between file formats. It took Apple less than forty-eight hours to take legal action against PlayFair.

A Software Bloom: Code War Momentum

PlayFair was released on the open-source Web site SourceForge.org on April 6, 2004. On April 8, Robin Miller, the copyright agent for SourceForge.net, received a cease-and-desist email and fax from David Hayes, a lawyer with the San Francisco firm Fenwic and West, the firm representing Apple Corporation. In that letter, Mr. Hayes, citing the DMCA's anticircumvention provisions, demanded that the SourceForge.org site take down the PlayFair project from its servers and cease distributing the PlayFair client. On April 9, SourceForge.org pulled the project from its servers. Not to be outdone, the developer of PlayFair then reposted the project on Sarovar.org, an open-source development site similar to SourceForge located in India. By April 15, however, the Sarovar.org administrators had also received a cease-and-desist letter from Apple through its legal representation in India. In that letter, Apple noted that Sarovar was in violation of India's Information

Technology Act of 2000 and Copyright Act of 1957. In response to this letter, Sarovar withdrew the PlayFair project from its servers (Sarovar 2004).

PlayFair spawned a long passionate discussion among iTunes loyalists and PlayFair supporters. The commentary is illuminating and addresses the frustrations users felt with the iTunes DRM system. It also illustrates that by no means are users monolithic in their feelings toward DRM. Some users felt that the FairPlay DRM system was as good a deal as could be had in the digital environment. Yet others felt that they were entitled to more. Two opposing views indicate the conflicts:

> I have the answer to all your problems: **Stop purchasing music with DRM!** You knew, before you purchased anything from the iTMS, that you would **not** be able to convert it to another format; you would **only** be allowed to play it on three computers at a time, plus iPods; and that you would **not** be allowed to remove the copy protection from the files you purchase. Quit whining about "legitimate" uses when you **agreed to give up** those options when you purchased from the iTMS. You agreed to a contract and are attempting to, or already have, violated that contract. **That** is **illegitimate**. (aogail 2004, emphasis in original)
>
> Moron . . . calling me a thief. . . . I wanted to make a music video in mpg format out of an itunes file and some anime. I was unable to because of "fairplay." Not so fair eh? Your shortsightedness is amusing, and you make it seem as if there is parity between intellectual and actual property. ("Re: Overseas" 2004)

These comments give us extraordinary insights into the complexities of the issues surrounding digital copyright at the level of users. The second user's understanding (or lack thereof) of fair use, "aogail's" understanding (or lack thereof) of market power in contract, and the power of technological resistance to make a reality out of a belief play themselves out in this exchange. Imagine what it would mean to understand fair use in this way without PlayFair. PlayFair makes the expanded conception of fair use real. The software expands the alternative conception through its distribution on the Internet by gaining users and by creating, through practice, a new fair use, regardless of the law. As I noted in the introduction to this book, the realization of a world where user conceptions are made reality is a central distinguishing factor of digital rights activism. PlayFair is an example of this factor. It is the ideal made real.

These computer programs are not trivial, and lest we doubt the undertones of civil disobedience in these technological practices, some users surely recognize them:

> People can argue ethics all they want. But the simple fact is that the music industry should make the experience the best they can for their legitimate customers, of which I am one. I have had numerous problems trying to play my songs on differ-

ent computers I'm using. I'm a technically savvy guy, and I know all I need to do is contact tech support. But you know what? Just because I own four computers, I don't feel like asking permission each time I want to play a song that I paid them for. Face it: DRM sucks. I don't want to steal. DRM still sucks. So I'll be using PlayFair to unlock my legally purchased songs, and then I'll enjoy them however I want. Yeah, it's illegal. But it's "right" in my book, and I'm not ashamed to flaunt [*sic*] these crappy laws. (Localman 2004)

In spite of Apple's persistence, PlayFair was back online by May 10, 2004, registered to an Indian citizen and hosted on Hymn-project.org with legal support from the Free Software Foundation India. According to its developer, the software was downloaded three hundred thousand times as of May 14, 2004. It also had acquired a new name, "Hymn," an acronym for "Hear Your Music Anywhere" (Hymn Project 2004e).

DeDRMS

On April 25, at the same time that PlayFair was confronting its problems with Apple, Jon Johansen released DeDRMS, an application quite similar to PlayFair. The DeDRMS utility worked on Windows machines in conjunction with VLC by using the song keys generated by drms.c to create new copies of a song without any DRM protection. Recall that VLC could only generate keys for protected songs that needed to be copied if a user wanted to move music to more than the three authorized computers. With DeDRMS, the user could use VLC to get the song keys from the computer system; DeDRMS would convert the songs into unprotected AAC files. The difference between drms.c and DeDRMS was that the latter generated unencrypted files that would not need copies of song keys to be played on other computers.

Because songs purchased on iTunes have included in their code information about the user who bought the songs, both Johansen and the PlayFair/Hymn developer (from this point, I refer to PlayFair as Hymn) designed their applications to incorporate a technological system of self-policing into the software. If a user chose to decrypt his or her iTunes songs and then share them on peer-to-peer networks, for example, the songs would be traceable back to the person who originally bought the song. Johansen makes his intent clear in the "readme" document that comes with the DeDRMS source code. He writes: "DeDRMS requires that you already have the user key file(s) for your files. The user key file(s) can be generated by playing your files with the VideoLAN Client. . . . DeDRMS does not remove the UserID, name and email address. The purpose of DeDRMS is to enable Fair Use, not facilitate copyright infringement" (Johansen 2004).

The Hymn developer also makes his[17] intentions clear in the "frequently asked questions" (FAQ) section of the Hymn user manual:

Q: Why are you trying to promote music "piracy"? Shouldn't musicians make money, too?
A: First of all, I buy all of my music. In fact, most of the music I buy these days comes from the iTunes Music Store. However, I want to be able to play the music I buy wherever I want to play it without quality loss, since I PAID FOR that quality. I want musicians to make money. I want Apple to make money. I don't condone sharing music through P2P [peer-to-peer] networks with the masses, though I believe making a mix CD or playlist for a friend is okay. I also think the RIAA are a bunch of crooks, but that's another story.

Secondly, hymn leaves the apple ID embedded in the output file, so anyone who shares the decoded files on P2P networks is bound to be prosecuted under copyright law. (Hymn Project 2004c, emphasis in original)

Once again these comments bear out some underlying themes in the copyright debate. They illustrate practices that are at odds with the law and the norms of technology and media use that conflict with fair use as it is legally defined. The spirit of this debate is encapsulated in the user's statement quoted previously: "Yeah, it's illegal. But it's 'right' in my book" (Localman 2004). Importantly, such comments also add a layer of complexity to any understanding of how hackers who facilitate a user-centered interpretation of fair use see that doctrine. For them, licensing that precludes users from having reasonable access to their cultural goods is limiting, but they themselves also define some limits. The digital rights movement's interpretation of access and use rights does not advocate a free-for-all, and the more moderate activists and leaders would like to see a compromise.

FairKeys

One of the ways Apple ensured that a user could play a song on only three authorized computers was to keep a copy of the song keys on its servers. The song keys were system specific, so if songs were installed on a computer that was not authorized (and so no song keys were found locally), the iTunes software would pull the keys from the iTMS server and attempt to open the songs with the keys. If the system information (as described in the section on drms.c) on the keys did not match the computer on which the songs are stored, iTunes would ask if the user wished to authorize the computer to play the songs. This question assumed the user had not yet authorized the three computers stipulated by the EULA. If the user had already authorized three computers, however, iTunes would not play the songs. Under normal circumstances, a user with three authorized computers who acquired a fourth computer that he or she wanted to play music on would be required to deauthorize one of the three originally authorized computers. iTunes would then communicate the deauthorization to the iTMS server, which would then update the server keys, noting that only two computers were authorized and that the user could now authorize a third computer.

Following upon the release of DeDRMS, Johansen released FairKeys, an application that would retrieve a user's song keys from the Apple iTMS servers. With this new tool, a user could access the Apple servers using her log-in information. The Apple server, believing that the user was accessing the songs with iTunes, would send the song keys to FairKeys. As explained earlier, iTunes would not have played the songs because the keys would note that three computers were already authorized. DeDRMS, however, could use those keys to strip the DRM from the protected song (see figure 7.9). The FairKeys/DeDRMS combination made it unnecessary to use VLC (something already done by Hymn) or to have the song keys on the system attempting to strip the DRM; those keys were now accessible via FairKeys' access to the iTMS servers.

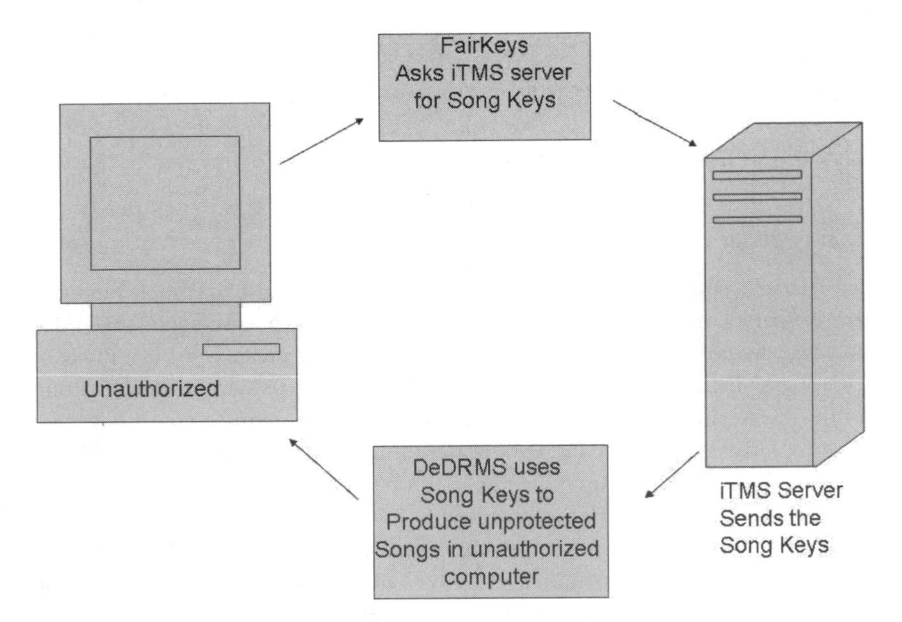

Figure 7.9
How FairKeys and DeDRMS produce unprotected songs.

FairKeys used the user's log-in information to get song keys from the iTMS, so the self-policing aspect of the software was preserved because, in theory, only a person who legally bought the songs would have the log-in information matching the copies of the songs in question.

iOpener and JHymn

The Hymn Project developer quickly noted FairKeys on the Hymn Project Web site. With this added feature, Mac users using Hymn did not need to have an iPod mounted on their desktops to be able to strip the DRM from songs bought on the iTMS.[18] Furthermore, from May 2004 until early 2005 the Hymn Project expanded, making it increasingly easy to strip DRM from iTunes songs. In August 2004, a Hymn Project community member released iOpener for Windows, a variant of Hymn, which ran on Microsoft's ".net" framework. iOpener monitored the iTunes music library on the installed computer and would convert DRM-protected iTunes files into unprotected AAC files, allowing for almost seamless integration (Hymn Project 2004d). On the iOpener FAQ, the developer stated:

> iOpener is an application that will find all of the "protected" AAC files in your iTunes library (the ones you purchased online) and remove the DRM (encryption) from them "in place," allowing you to enjoy the music you've purchased on any device anywhere that supports the standard AAC format. This means that you will notice no change whatsoever in iTunes except that the "type" of the track will change from "Protected AAC audio file" to "AAC audio file." Additionally, iOpener can run in the background (in your task tray, actually) and auto-decrypt any "protected" AAC files as they are added to your iTunes library. (Hymn Project 2004d)

Regarding the legal implications of this software, the developer noted that iOpener is

> an insurance policy, primarily. In a capitalist economy like the one Apple (and you and I) operate in, it's not unlikely that Apple, Inc. will be bought/sold/merged/dissolved sometime between now and the day I'm pushing up daisies. I like the convenience of being able to purchase music online, but don't like the idea of sitting in the retirement home with a hard drive full of music from my younger days that I can't "authorize" because CocaNikeEnron, Inc. decided to shut down its "Apple division" to boost this quarter's figures. . . . Don't share music or other files on P2P [peer-to-peer] networks or elsewhere online unless you wrote/created them. It's unethical. . . . Making mix tapes for your friends (i.e. people whose faces or at least real names you know) is cool. Anonymously sharing your entire collection with the world is not. That being said, iOpener preserves all metadata . . . in the track. This includes your Apple ID and email address. Share if you want, but caveat emptor. (Hymn Project 2004d)

Also during this time, a Hymn community member who went by the online name "FutureProof" developed JHymn, another variant of Hymn but written in Java. This variant, originally written for the Mac, grew into the most user-friendly iTunes DRM system–stripping technology since the iTunes battle in the copyright code war began. JHymn had the ability to easily find protected files and convert them into unprotected AAC files. It gave the user the option of converting those AAC files to mp3 files "on the fly," and *it would strip the identifying information from the generated files*.

Apple responded to all these hacks by updating iTunes with features that would undermine circumvention measures. For example, iTunes 4.6 included a function that would search out unprotected AAC songs bought on the iTMS (with the identifying metadata still attached) and refuse to play them, going as far as deleting copies of these songs found on users' iPods (FutureProof 2004).

According to the developer's comments on the JHymn Web site, the user data on the unprotected songs was a "dead giveaway that those files once lived a life as DRM-protected files" (FutureProof 2004). Faced with the difficulty of having unprotected AAC files remain unplayable to the iTunes software (although they can be played by other applications, such as Winamp), the developer included a "scrubbing" option in subsequent versions of JHymn. The developers noted his discomfort in including the option to turn off the self-policing feature that had been a part of all the variants originating from PlayFair. He noted,

> For various philosophical and quasi-legal reasons, JHymn, right "out of the box" so to speak, does not strip your Apple ID or the copyright information contained within the files from which DRM is removed. . . . We know after what happened when iTunes 4.6 came out that such vulnerabilities can, and probably will, be exploited. . . . You have to make the decision yourself, as the user of JHymn, to remove either or both of these items by going into JHymn's Preference settings. (FutureProof 2004)

At play in this cautionary statement were the developer's commitment to what he believed to be fair use, his understanding of his liability should he have stripped the user information as a default, and his desire to continue to provide a tool that would effectively free iTunes songs from DRM.

It would be naive to think that the technological self-policing in some of these applications was conceived solely out of concern for preventing illegal music sharing on peer-to-peer networks. It is quite probable that self-policing features were included to diminish the appearance that such technologies were being developed with the explicit purpose of facilitating copyright infringement by allowing once-protected songs to be sharable.

Regardless, the self-policing features had an expressive function beyond that of their operative function. Technological self-policing features state, "Although we believe in expanded fair use, we do not condone rampant copying." Self-policing drew technologically defined boundaries around an ideology of fair use. It was, in a sense, a compromise on the part of the developers of these technologies. It expressed recognition of the damage that the technology might do to copyright owners and attempted to address it. Apple was unfortunately unwilling to entertain this discussion because it did not own the copyrights to the songs the owners sell.

Although statements about the recommended use of these features were articulated to limit the distributor's liability, I do not believe that the self-policing features were present entirely for this reason. The very design and distribution of these programs was in violation of the various anticircumvention provision of DMCA section 1201. Therefore, the developers of these technologies were already in violation of the law by the very fact that they made and shared these programs.[19] Statements that attempt to mitigate the impact of these technologies for copyright were issued after the fact, and it would be folly for a developer to rely on a judge or jury to take them as genuine. As such, statements about intended use must have some value other than buffering liability—that value being an honest expression of a belief both in copyright and in its limitations.

Conclusion

All told, between April 2003 and the summer of 2005, iTunes faced twenty-four services and technologies designed with the explicit purpose of circumventing the restrictions on copying and accessing songs bought on the iTMS (table 7.1).

Some of these technologies were designed with interoperability in mind (the ability to play songs on more than just iTunes); many were open-source projects. The most successful were designed with the explicit political purpose of giving users of iTunes what designers felt were fair-use rights. Hymn stands out among these. The developer had very explicit viewpoints on the matter of Hymn's being a political technology that allowed users their rights:

> The purpose of the Hymn Project is to allow you to **exercise your fair-use rights** under copyright law. The various software provided on this web site allows you to free your iTunes Music Store purchases (protected AAC/.m4p) from their DRM restrictions *with no loss of sound quality*. These songs can then be played outside of the iTunes environment, even on operating systems not supported by iTunes and

Table 7.1
List of iTunes Hacks from March 2003 to Summer 2005

iTunes Hacks	Release Dates
iCommune (for Mac)	March 2003
ServerStore	April 30, 2003
iSlurp	May 12, 2003
iLeech	March 2003
iTunesDL	May 17, 2003
iTunesTracker	June 2003
MyTunes (for Windows)	October 26, 2003
QTFairUse	November 17, 2003
drms.c	January 5, 2004
VLC 0.7.1 Functionality	March 2, 2004
PlayFair (for Windows and Mac)	April 6, 2004
Hymn	April 25, 2004
AppleRecords	May 19, 2004
JHymn	Summer 2004
DeDRMS	April 25, 2004
FairKeys	July 7, 2004
iOpener	August 22, 2004
FairTunes (for Mac)	April 25, 2004
OurTunes	August 9, 2004
GetTunes (Mac version of MyTunes)	Summer 2004
SharpMusique	March 31, 2005
PyMusique	March 18, 2005
JusteTune	April 15, 2005
Musik	Summer 2005

on hardware not supported by Apple. . . . "The primary objective of copyright is not to reward the labor of authors, but [t]o promote the Progress of Science and useful Arts." "To this end, copyright assures authors the right to their original expression, but encourages others to build freely upon the ideas and information conveyed by a work. This result is neither unfair nor unfortunate. It is the means by which copyright advances the progress of science and art."—**US Supreme Court Justice Sandra Day O'Connor**. Despite what Justice O'Connor of the United States Supreme Court has said, DRM exists. The purpose of DRM is to bypass traditional copyright law. The result of DRM combined with laws that outlaw circumvention of DRM (such as the DMCA) is that there is no longer clear protection for fair use in some countries. (Hymn Project 2004a)

Also of note is the fact that these circumvention technologies became increasingly easier to use. The evolution in usability from the initial

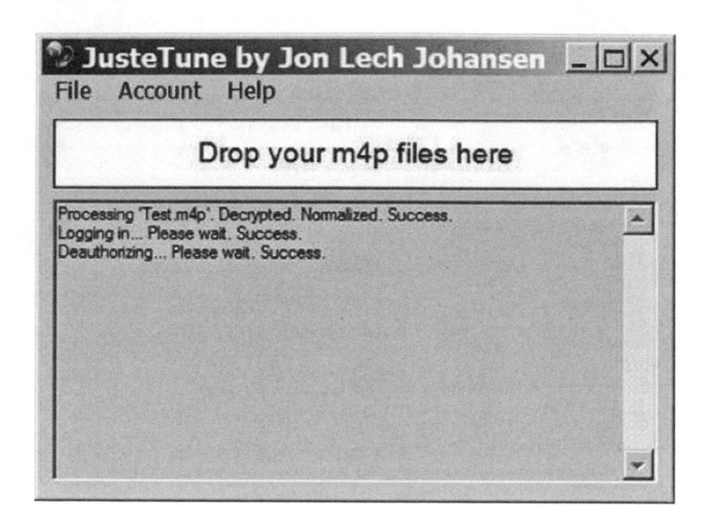

Figure 7.10
Drag-and-drop function of Jon Johansen's JusteTune iTunes hack.

command-line interface to drag-and-drop functionality contributed to wide adoption of these technologies (see figure 7.10).

These iTunes DRM-circumvention technologies distinguish themselves from the AEBPR and DeCSS in that they were explicitly designed to challenge technological enforcement that designers felt embodied the DMCA's overly broad limitations. Unlike the latter two technologies, iTunes hacks did not gain a political meaning based on struggles in court but were instead born with that meaning; their adoption, although convenient for many, was a matter of politics for others (especially the designers).

Apple eventually abandoned the DRM system for music sold on the iTMS, but the way technologies like them and the hacks to them were understood by corporations and the movement reflect an important cultural framing of technology as part of the ongoing debate over user rights. Technology itself is both a boon and a hindrance to participation, and both sides of the debate claim technology for their own purposes.

On the side of those companies selling digital music (such as Apple), digital technologies were presented as untethering the user from the confines of old media and allowing him or her to roam creatively through music collections. For example, the now famous ad campaigns for iPods showing silhouetted figures dancing ecstatically intimated a liberated user, faceless so that he or she could be any of us, but joyous, unattached to any physical world save the iPod colored in its signature white. This image, coupled with Apple's early ad campaign with the tagline "It's your music . . . rip, mix, burn," sent a clear message: digital music and digital technology would free you to consume music in a personal way.

The politics of these technologies, then, are related not only to legal debates (the legal politics of copyright), but also to the cultural politics of music consumption, cultural appropriation, and ownership. The ad campaign added a certain type of rhetoric about digital technologies; it presented an argument for freedom. Of course, the great irony was that this freedom was not exactly what users encountered. The Apple DRM system tied users to iPods, the iTMS, and iTunes. Far from liberating a user from old media and technologies, they tied them to new ones.

It is not surprising that the digital rights movement itself generated its own rhetorical moves—through, for example, naming one of the hacks to the iTunes DRM system "PlayFair." These semantic hacks were and continue to be important to the movement. They give cultural meaning to politics that reside deeply in the workings of technological systems. Because they do so, users can connect to a particular technological system not only functionally but also meaningfully. This practice common to highly politicized digital technologies can be seen in, for example, studies of user attitudes toward adoption of open-source software (such as the Mozilla Firefox browser): although users adopt the technology for functional purposes, they also do so because of its meaningful politics.

The irony represented by Apple's "rip, mix, burn" mantra and its implementation of DRM ultimately suggests that corporations selling digital music often find themselves in a middle ground between copyright owners' desire to strictly control consumption and users' strong participatory impetus. On which side companies such as Apple fall is a matter of convenience. Apple supported DRM because it needed to appease copyright owners, but it jettisoned DRM when the system became intractable, its competition started offering DRM-free music (services such as Amazon .com never used DRM), and users rejected DRM.

Chapter 8 further discusses the implications of the role of technology as a means of making a political statement and as a means of affecting change with both its functional aspects and its meaning. Hackers such as FutureProof and Johansen were not the only ones producing such explicitly political technological resistance. Other technologies were also produced, targeting both the iTunes DRM system and digital copyright in general. These technologies were developed as an explicit tactic employed by active individuals and movement organizations.

8 Structure and Tactics of the Digital Rights Movement

This chapter documents the tactics of key SMOs and actors in the digital rights movement through brief case studies. Here I seek to map the actors in the field and position technological resistance within the movement as a whole. Importantly, as this chapter reviews the movement's tactics, it also positions technology in a way that may decenter SMOs as the *key players* in mobilization. This observation is significant in the context of social movement studies, which typically position organizations as central for both framing movement issues and mobilizing resources and adherents for collective action. I do not imply that the SMOs are irrelevant in the digital rights movement—they remain a crucial component—but rather that hackers and their wares are new and important players that can shift and have shifted movement dynamics. The fact that technological resistance is a very important tactic in the movement repertoire is illustrated by the myriad protest and mobilization strategies presented in the case of each organization. These organizations, like organizations in other social movements, play to their strengths, adopting tactics that suit their members' skill set and resources (McAdam, McCarthy, and Zald 1996).

Digital Rights Movement Organizations and Their Tactics

As of 2011, there were twenty-five identified organizations that actively engage in advocating the issues of the digital rights movement (table 8.1), many of which focus on issues of access and use of copyrighted material. (Appendix C lists these organizations and their mission statements.)

Not all organizations that work on issues pertinent to the digital rights movement are presented in table 8.1, and of those included, I discuss only the ones that have shown through research to be exemplars of theoretical points particular to the movement and that stand out as leaders.

Table 8.1
Digital Rights Movement Organizations and Classification

Organization	Class
Creative Commons	Nongovernmental Organization (NGO)
Free Software Foundation	NGO
Samuelson Law, Technology, and Public Policy Clinic, Berkeley Boalt School of Law	law clinic
EFF	NGO
Global Internet Liberty Campaign	NGO
Lawrence Lessig Blog	blog
Electronic Privacy Information Center	NGO
Copyfight	blog
Berkman Center for Internet and Society, Harvard Law School	legal center
Center for Democracy and Technology	NGO
Public Knowledge	NGO
Center for Internet and Society, Stanford Law School	NGO
Chilling Effects: Cease-and-Desist Clearinghouse	NGO
Computer Professionals for Social Responsibility	NGO
American Libraries Association	NGO
Downhill Battle	NGO
Free Culture	grassroots organization in transition to NGO
Privacy Rights	NGO
Digital Future Coalition	NGO
Participatory Culture Foundation/Get Democracy	NGO
Future of Music	NGO
Our Media	NGO–grassroots organization
America Association of Law Libraries	NGO
Homer Recording Rights Coalition	industry–consumer coalition
Association for Computing Machinery	NGO

The tactical repertoire of a movement—the tool set that a social movement uses to further its cause—may include a host of strategies involving (1) choice in organizational structure (grass roots or social movement or both); (2) types of mobilization undertaken (institutional or extrainstitutional); (3) framing strategies that try to capture the meanings of the social movement (e.g., media capture); and (4) technological design (as in iTunes hacks). This chapter describes three SMOs—the EFF, Creative Commons, and Downhill Battle—which together represent the full spectrum of tactics used in the digital rights movement.

The EFF

The EFF fits the traditional definition of an SMO. It relies on its constituency for support and engages in both institutional and extrainstitutional tactics to further movement goals.

The EFF broadly engages the issues of the digital rights movement, focusing not only on copyright issues, but also on privacy and consumer rights in technology and entertainment products; free-speech issues such as blogger's rights, censorship online, and filtering systems; privacy issues such as biometrics and expansion of the Communications Assistance for Law Enforcement Act (CALEA) of 1994;[1] and intellectual property in digital networks and other digital technologies (see table 8.2). The EFF has confronted a number of intellectual-property issues, including: expanded trademark powers, expanded global intellectual-property regimes through multilateral treaties, and copyright issues both at home and abroad.

The EFF spends much of its energies in mustering support for laws that expand the digital rights cause. It has historically attempted to organize action against legislative initiatives it has perceived as a threat to digital rights. A list of some legislation that concerns the EFF is presented in table 8.3.

Commensurate with its issue campaigns, the EFF has supported legislation that would expand user rights in digital media currently protected by copy-control technologies (such as the Digital Media Consumers' Rights Act, still pending). It has set up prewritten form letters that can be e-mailed to representatives who support legislation that is unpopular within the movement (figure 8.1), and maintains an updated "Action Alert" section on its Web site with information about pending legislation that can affect its constituency. The EFF has done most of its work in the courts, taking on defense roles in the Sklyarov case, the DeCSS cases, and others not discussed in this work (such as Felten et al. v. RIAA, US District Court [NJ 2001]).

The EFF's role in landmark court cases in the movement has been pivotal in rallying support for the movement, defining the frames of the cases, and providing legal support. As noted earlier, in the Sklyarov case the EFF proved crucial in negotiations with Adobe, leading to Adobe's public statements against the Department of Justice even though Adobe had prompted it to initiate the investigation and provided important information leading to Sklyarov's arrest. Furthermore, the EFF served as a focal point for distributing information, linking groups that were actively protesting Adobe, helping to organize a defense fund, and providing initial legal support. The continuous press releases by the EFF in the Sklyarov case also helped define the case in terms of free speech and freedom of

Table 8.2
List of Issues Covered by the EFF

Issue	Advocacy
Anonymity	Advocacy against regulation that would take away anonymous communication over digital networks
Biometrics	Advocacy against the deployment of biometric technology that might infringe on privacy and discriminate
Bloggers' rights	Advocacy of journalistic protection of free speech for bloggers
Broadcast flag	Advocacy against transmission of a security signal on television content that would dictate copyright management for digital receivers such as TiVo or digital televisions
CALEA	Advocacy against the expansion of the CALEA, which would allow for easier wire tapping of Internet communications
CAPPS II	Advocacy against the Computer Assisted Passenger Prescreening System (CAPPS) by the Transportation Security Administration, a profiling program that would require airline passengers to surrender personal information
Censorship	Advocacy against censorship of all manner of speech (commercial and noncommercial) online
Copyright law	Advocacy of a broad set of copyright freedoms, including limitations on DRM technology, curtailment of the DMCA, freedom with domain names, and protection of file-sharing technologies, with the main goal to increase creative freedom, freedom to innovate, consumer rights, free speech, and academic freedom
E-voting	Advocacy for openness in e-voting technology, implementation of fail-safes and research
Filtering	Advocacy against browser-filtering systems implemented in public-use terminals such as libraries
International trade	Advocacy against the Free Trade Area of the Americas legislative initiative that would require signatory nations to implement anticircumvention provisions like the DMCA's section 1201
Internet governance	Advocacy for core issues such as privacy and free speech in the Internet Corporation for Assigned Names and Numbers regulation
ISP legalities	Advocacy for limited liability for ISPs in cases of intellectual-property violation

Table 8.3
Partial List of Pending Laws That the EFF Has Lobbied For or Against

Legislation	EFF's Action
Audio Broadcast Flag Licensing Act of 2006 and the Digital Content Protection Act of 2006	Fights against a mandate on digital radio and TV manufacturers to place technology that would implement audio signal security limiting copying functions for specific broadcast content.
The National Security Agency's Domestic Surveillance Program	Fights for strict investigation of the National Security Agency's surveillance program.
Digital Transition Content Security Act (HR 4569, "Analog Hole bill")	Fights against placement of watermarking technology on all audio/visual media players to prevent postprocessing capture of content.
WIPO's Treaty on the Protection of Broadcasting Organizations	Fights against giving copyrights to broadcasters even if the content is openly licensed by the creators.
Radio Frequency Identification tag legislation in California	Fights against Radio Frequency Identification tags in state identification cards such as drivers' licenses.
Voter Confidence and Increased Accessibility Act	"Fights for mandate requiring a paper audit trail for all electronic voting machines, random audits, and public availability of all code used in elections" (EFF 2005).
Digital Media Consumers' Rights Act	Fights for legislation that "would give citizens the right to circumvent copy-protection measures as long as what they're doing is otherwise legal" (EFF 2005).
National Weather Services Duties Act	Lobbies against legislation that would limit the National Weather Service's release of weather data to the public when those data compete with commercial interests.
Trademark Dilution Revision Act	Lobbies against legislation that would expand trademark powers and increase the liability of supposed infringers.

innovation. For example, almost immediately after Sklyarov's arrest the EFF mobilized protest and framed the case to its advantage, as in the following press release that helped organize protest against Adobe:

The Electronic Frontier Foundation and community activists urge concerned citizens to join in a San Francisco Bay Area protest on Monday, July 23, against software firm Adobe's role in the jailing of programmer Dmitry Sklyarov. . . . Adobe, seeking to protect electronic property rights at any cost, has apparently pushed the U.S. Department of Justice into an ill-advised arrest of a Russian programmer under the Digital Millennium Copyright Act. . . . Our hearts go out to Dmitry's wife, children,

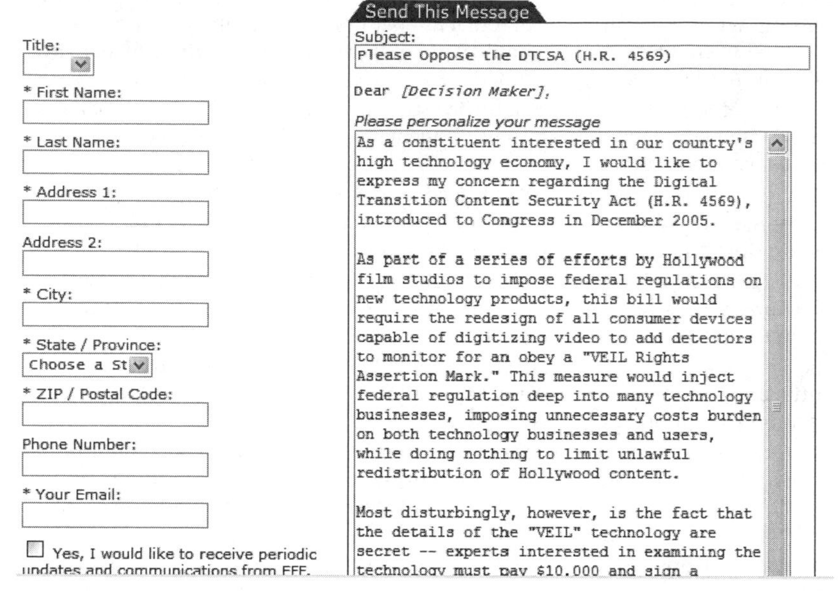
1. Complete the form below with your information.
2. Personalize the subject and text of the message on the right with your own words, if you wish.
3. Click the Next Step button to send your letter to these decision makers:

- Your Representative

Send This Message

Title:

* First Name:

* Last Name:

* Address 1:

Address 2:

* City:

* State / Province: Choose a St

* ZIP / Postal Code:

Phone Number:

* Your Email:

Yes, I would like to receive periodic updates and communications from EFF.

Subject:
Please Oppose the DTCSA (H.R. 4569)

Dear *[Decision Maker]*,

Please personalize your message

As a constituent interested in our country's high technology economy, I would like to express my concern regarding the Digital Transition Content Security Act (H.R. 4569), introduced to Congress in December 2005.

As part of a series of efforts by Hollywood film studios to impose federal regulations on new technology products, this bill would require the redesign of all consumer devices capable of digitizing video to add detectors to monitor for an obey a "VEIL Rights Assertion Mark." This measure would inject federal regulation deep into many technology businesses, imposing unnecessary costs burden on both technology businesses and users, while doing nothing to limit unlawful redistribution of Hollywood content.

Most disturbingly, however, is the fact that the details of the "VEIL" technology are secret -- experts interested in examining the technology must pay $10,000 and sign a

Figure 8.1
Image of a form letter that can be directly emailed to a congressperson opposing the Digital Transition Content Security Act (HR 4569).

and colleagues who are likely distraught by what appears to be a most disgraceful arrest. . . . We protest Adobe's role in perpetrating this grave miscarriage of justice. . . . The San Francisco Bay Area protest will occur at Adobe Headquarters, from 11:00 am to 1:00 pm this Monday, July 23. Protestors will gather in San Jose at the Quetzalcoatl snake sculpture at the south end of Cesar de Chavez Park, at the corner of South Market St. and West San Carlos St, then march to Adobe Headquarters at 345 Park Avenue in San Jose. . . . Protest organizers include the Electronic Frontier Foundation, BoycottAdobe.com, and a loose-knit group of activists linked together through the free-sklyarov email list. The organizers request that attendees bring along U.S. or Russian flags and signs. Free T-shirts from a group called BoycottAdobe.com will be distributed to the first fifty attendees. (EFF 2001d)

The EFF also has adopted a strategy of providing its constituency with as much primary documentation as possible. On its Web site, for example, it dedicated a whole section to the Sklyarov case, making available deposition transcripts, transcripts of testimonies, and transcripts of hear-

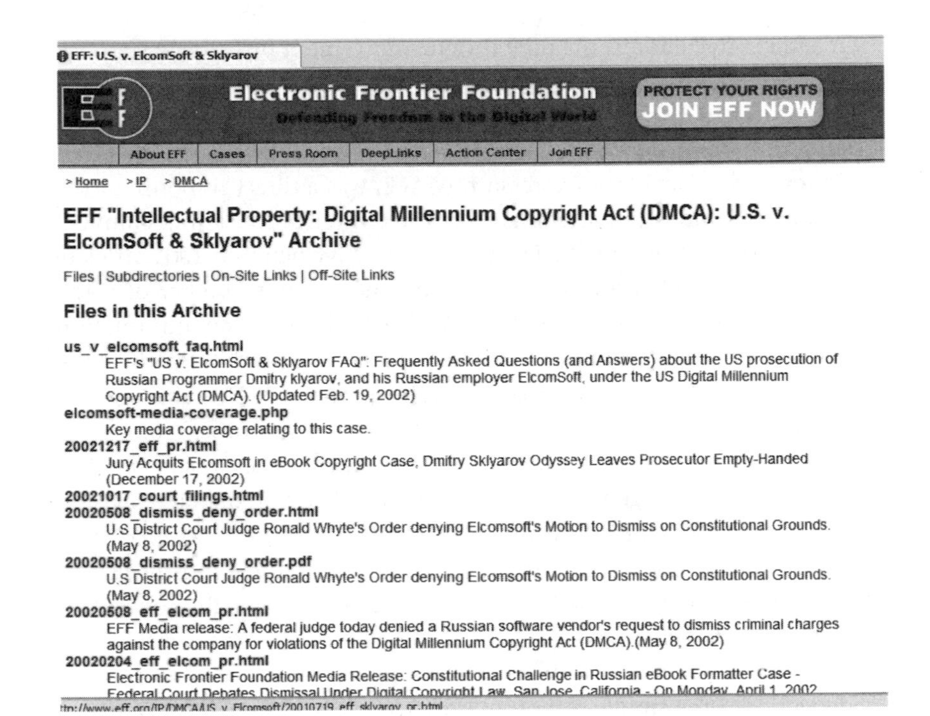
EFF: U.S. v. ElcomSoft & Sklyarov

Electronic Frontier Foundation

PROTECT YOUR RIGHTS
JOIN EFF NOW

About EFF | Cases | Press Room | DeepLinks | Action Center | Join EFF

> Home > IP > DMCA

EFF "Intellectual Property: Digital Millennium Copyright Act (DMCA): U.S. v. ElcomSoft & Sklyarov" Archive

Files | Subdirectories | On-Site Links | Off-Site Links

Files in this Archive

us_v_elcomsoft_faq.html
EFF's "US v. ElcomSoft & Sklyarov FAQ": Frequently Asked Questions (and Answers) about the US prosecution of Russian Programmer Dmitry klyarov, and his Russian employer ElcomSoft, under the US Digital Millennium Copyright Act (DMCA). (Updated Feb. 19, 2002)
elcomsoft-media-coverage.php
Key media coverage relating to this case.
20021217_eff_pr.html
Jury Acquits Elcomsoft in eBook Copyright Case, Dmitry Sklyarov Odyssey Leaves Prosecutor Empty-Handed (December 17, 2002)
20021017_court_filings.html
20020508_dismiss_deny_order.html
U.S District Court Judge Ronald Whyte's Order denying Elcomsoft's Motion to Dismiss on Constitutional Grounds. (May 8, 2002)
20020508_dismiss_deny_order.pdf
U.S District Court Judge Ronald Whyte's Order denying Elcomsoft's Motion to Dismiss on Constitutional Grounds. (May 8, 2002)
20020508_eff_elcom_pr.html
EFF Media release: A federal judge today denied a Russian software vendor's request to dismiss criminal charges against the company for violations of the Digital Millennium Copyright Act (DMCA).(May 8, 2002)
20020204_eff_elcom_pr.html
Electronic Frontier Foundation Media Release: Constitutional Challenge in Russian eBook Formatter Case - Federal Court Debates Dismissal Under Digital Copyright Law, San Jose, California - On Monday, April 1, 2002

Figure 8.2
Archive of DVD CCA v. Sklyarov documents on the EFF Web site. From EFF 2001a.

ings, briefs, and court orders almost instantaneously, usually within twenty-four hours after they were issued. For example, between December 2001 and December 2002, the EFF made available more than twenty-one primary documents from the Sklyarov case in a large archive that was easily accessible (EFF 2001a) (see figure 8.2).

This level of transparency gave the EFF a substantial amount of credibility because its press releases could be read against the reality of the court proceedings and judged on those merits. This transparency became increasingly important in the DeCSS case, where the MPAA fought to keep the proceedings secret.

In both DeCSS cases, the EFF took advantage of its ability to distribute information and provide links to primary documents. Also, by leading the defense, the EFF was able to gain some important allies in the fight against the DMCA. As noted in tables 6.1 and 6.2, many of the most prominent scholars in the field of digital copyright spoke out in favor of DeCSS. Thus, although the EFF had a mixture of success in these court cases, it positioned itself as a central actor in the digital rights movement by actively supporting the defendants.

Following the Sklyarov, DeCSS, and other cases, the EFF organized a mobilization campaign that involved activism at many levels. Its Campaign for Audiovisual Free Expression (CAFÉ) was a direct response to the entertainment industry's attempts to enforce the DMCA's anticircumvention provisions. During the CAFÉ, which was active between 2001 and late 2003, the EFF sought to mobilize its constituency in a number of ways. First, the CAFÉ section of the EFF's Web site acted as an information resource, collecting headlines regarding important issues that involved the DMCA, such as the RIAA's campaign against peer-to-peer file sharing, the use of the DMCA to block information on e-voting machines from being distributed online, awareness-raising notices over changes in contract law that would allow stringent licensing in software, DRM on digital television recorders, and other anti-DRM topics.

The campaign also coordinated letter-writing initiatives, mostly against unpopular bills that curtailed user rights over legally bought content or against legislation that prevented the development of circumvention and distribution technologies with legitimate uses. As noted earlier, the EFF made letter writing simple by automating the process though its Web site. Its approach to achieving social change was comparatively traditional. Working within institutional frameworks, it gathered allies to its cause (prominent technologists and law professors, for example) and worked within the courts and the legislature to change policy in digital copyright.

Despite its institutional approach, the EFF did use some extrainstitutional tactics, such as the protests organized during the Sklyarov case. Also, it attempted to define the frames in the fight over digital rights in music. The EFF and its supporters, for example, actively countered the RIAA's pamphletting in Los Angeles and New York with their own pamphlets and bumper stickers (figure 8.3). In an act of what activists called "guerrilla remixing," EFF supporters placed the bumper stickers illustrated in figure 8.4 over the RIAA's own pamphlets (which read, "Feed a musician, download legally"), crystallizing an important theoretical point in social movement theory: meaning matters, and contests over meaning are important for cohesion and continuance of the movement.

Although it is probably the case that the EFF did not coordinate this contest on the walls of the streets of US cities, it did do much to create and foster the user-centered meanings assigned to fair use and digital rights. Other organizations in the digital rights movement also engaged in this sort of tactic and took it to much greater levels.

Figure 8.3
Pamphlet released by the EFF to counter the DMCA and RIAA's campaign against peer-to-peer technologies.

Figure 8.4
Image of the RIAA's pamphlet with the EFF's sticker over part of the slogan. From http://www.boingboing.net/2005/08/10 /riaa_street_campaign.html.

Creative Commons: Restructuring the Legal Structure

One of the most important developments in the digital rights movement was the founding of Creative Commons in 2001 by Lawrence Lessig and students and fellows of the Berkman Center for Internet and Society at Harvard Law School and the Center for Internet and Society at Stanford. Creative Commons has for the most part stayed out of the courts (although Lessig has argued some important cases such as Eldred v. Ashcroft [537 US 186] in 2003). Also, as noted earlier, SMOs typically play to their strengths, and in this case Creative Commons has used legal expertise differently than the EFF.

Creative Commons deploys an ingenious tactic against the restrictive elements of copyright in digital media. Rather than undermine copyright law outright, it has devised a set of licenses that allow artists to give consumers certain rights as a default while reserving some rights for themselves. In a framing strategy intended to capture the social and legal purchase of the copyright symbol ©, it came up with its own symbol for its licenses (see figure 8.5). It also designed various licenses that can be offered to the consumer, the six most popular of which are listed in table 8.4. Its tactic is important because it captures the licensing convention that has served copyright owners so well. By using contract and flexible licenses, Creative Commons presents copyright owners with the possibility of maintaining more open contractual relationships with consumers. Its licensing schemes reenvision the consumer as someone who may want to take an active part in the creative process. Much like hackers working on technological resistance, the licensing scheme has layered functions. It serves a symbolic function by challenging the emblems of copyright ownership and putting alternative spins on well-known visual images (for example, the backward © to illustrate the reverse of copyright, a reversal borrowed from the open-source movement). It also serves a functional purpose by creating legally binding contracts that will enjoy the same protection as other commercial contracts and a structural purpose by creating an

Figure 8.5
Creative Commons symbol. From Creative Commons 2005.

Table 8.4
Examples of Creative Commons Licenses

Symbols	Name of License	What It permits
	Attribution Non-Commercial No Derivatives (by-nc-nd)	"Allows others to download works and sharing with others as long as they mention you and link back to you, but they can't change them in any way or use them commercially."
	Attribution Non-Commercial Share Alike (by-nc-sa)	"This license lets others remix, tweak, and build upon your work non-commercially, as long as they credit you and license their new creations under the identical terms. Others can download and redistribute your work just like the by-nc-nd license, but they can also translate, make remixes, and produce new stories based on your work. All new work based on yours will carry the same license, so any derivatives will also be non-commercial in nature."
	Attribution Non-Commercial (by-nc)	"This license lets others remix, tweak, and build upon your work non-commercially, and although their new works must also acknowledge you and be non-commercial, they don't have to license their derivative works on the same terms."
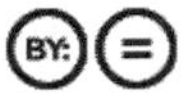	Attribution No Derivatives (by-nd)	"This license allows for redistribution, commercial and non-commercial, as long as it is passed along unchanged and in whole, with credit to you."
	Attribution Share Alike (by-sa)	"This license lets others remix, tweak, and build upon your work even for commercial reasons, as long as they credit you and license their new creations under the identical terms. This license is often compared to open source software licenses. All new works based on yours will carry the same license, so any derivatives will also allow commercial use."
	Attribution (by)	"This license lets others distribute, remix, tweak, and build upon your work, even commercially, as long as they credit you for the original creation. This is the most accommodating of licenses offered, in terms of what others can do with your works licensed under Attribution."

Source: Quotes from Creative Commons 2005.

ownership scheme parallel to copyright. This new contract with the consumer creates a user–creator relationship that undermines typical conceptions of how consumers and creators should interact. In the case of Creative Commons, success is gauged by adoption. It would serve the movement little to propose a licensing scheme that no one uses, just as it would serve the movement little to design technologies that no one uses. But Creative Commons has enjoyed enormous success; writers such as Cory Doctorow, Eric Raymond, and Lawrence Lessig have published works using Creative Commons licenses. Lessig wrote the second edition of his popular book *Code* (1999) as a wiki project under a Creative Commons license.

The project is a testament to social computing, participatory culture, and flexible, user-centered copyright. Universities have released educational content under Creative Commons licenses, and bloggers have widely adopted Creative Commons licenses for their content. Furthermore, digital publishing software applications such as Adobe Acrobat now include an authoring tool that allows the writer to set Creative Commons permissions on documents generated with Adobe. Within a year of the founding of Creative Commons, more than one million products were licensed with its license. The organization has started work on projects such as iCommons, an initiative to export the Creative Commons licenses to countries outside the United States, and Science Commons, an initiative to bring Creative Commons licensing to scientific publications and work.[2]

Whereas Creative Commons has attempted to capture the legal structure for the movement's purposes, other organizations and individuals hope to capture its technological structure through extrainstitutional tactics and hacking. The organization Downhill Battle is such case.

Downhill Battle, Grey Tuesday, and the Battle Labs

Downhill Battle was a not-for-profit organization founded in 2003 and based out of Worcester, Massachusetts (it is no longer active). It engaged a variety of issues working toward widespread or decentralized cultural production. It encouraged, for example, the use of peer-to-peer technologies to foster the distribution of independent artists' music to help them remain outside major media control. Its mission statement read:

> Five major record labels have a monopoly that's bad for musicians and music culture, but now we have an opportunity to change that. We can use tools like filesharing to strengthen independent labels and end the major label monopoly.
>
> How do musicians get paid for downloads? Simple: collective licensing lets people download unlimited music for a flat monthly fee ($5–$10) and the money goes to

musicians and labels according to popularity. This solution preserves the cultural benefits of p2p, gets musicians way more money, and levels the playing field.

Our plan is to explain how the majors really work, develop software to make filesharing stronger, rally public support for a legal p2p compensation system, and connect independent music scenes with the free culture movement. (Downhill Battle 2005a)

Its activities, however, went beyond the use of peer-to-peer technologies to achieve a "free culture" whose central tenet was access. Downhill Battle coordinated what it called acts of "digital civil disobedience" and developed technologies to counter the technolegal order established by digital copyright.

Digital Civil Disobedience: Grey Tuesday

During two weeks in December 2003, DJ Brian Burton (known as Danger Mouse) holed up in his Los Angeles home and produced one of the most critically acclaimed hip-hop albums of 2004. *The Grey Album,* as Burton called it, brought together the vocals of hip-hop artist Jay-Z's *Black Album* with the music from the Beatles' *White Album.* The fusion of the Beatles' pop and the hard core rap lyrics from Jay-Z melded two otherwise disparate genres into a coherent thread that was at once novel and familiar. It made widely evident the power of both digital technology and participatory culture, prompting one fan quoted in the *New York Times* to note, "[T]o a lot of artists and bedroom D.J.'s, who are now able to easily edit and remix digital files of their favorite songs using inexpensive computers and software, pop music has become source material for sonic collages" (Werde 2004).

Burton sent the tracks to some of his friends, and in a matter of days *The Grey Album* was a heavily sought-after album. Burton had about three thousand CD copies of the album made for promotion and distributed some of them over a period of a month. By the beginning of February, copies had made their way onto e-Bay, selling for as much as eighty dollars per CD. They were also found on the shelves of indy hip-hop music shops. The album even had cover art designed to show its sampling roots (see figure 8.6).

Once the album became so well known, EMI, which controls the Beatles catalog (the owners of the copyrights on the Beatles catalog are Sony and Michael Jackson's estate), sent Burton a cease-and-desist letter. Burton immediately pulled the album files from his Web site (he had also been distributing them online for free) and stopped giving out the promotional demos. But in Internet time it was too late. When news of EMI's cease-and-desist

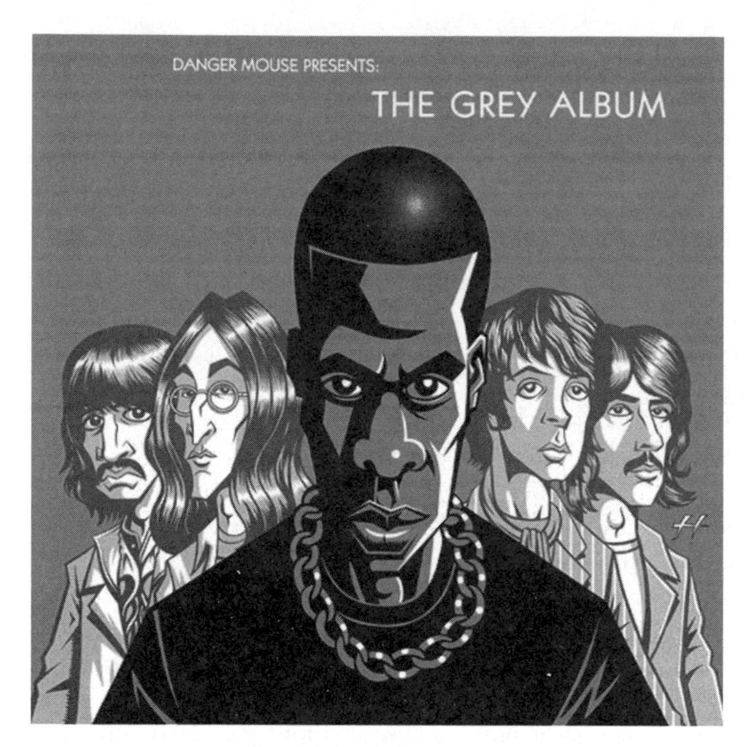

Figure 8.6
Cover of Brian Burton's *The Grey Album*. Design by Justin Hampton.

letter was reported on *Wired* and other news services, *The Grey Album* became a hot commodity on peer-to-peer networks, and activist groups such as Downhill Battle started to voice their objections.

At the heart of their complaints was an understanding that copyright law had swung too far in the direction of copyright owners. Activists argued that if copyright was supposed to foster further creativity by granting only limited monopolies, then ever-expanding copyright terms and restrictive regulation had now defeated the very purpose of its design. Jonathan Zittrain of the Berkman Center for Internet and Society at Harvard Law School agreed with this critique, noting that

> [c]opyright law was written with a particular form of industry in mind. The flourishing of information technology gives amateurs and home-recording artists powerful tools to build and share interesting, transformative, and socially valuable art drawn from pieces of popular culture. There's no place to plug such an important cultural sea change into the current legal regime. (Zittrain quoted in Werde 2004)

Within a few days of EMI's cease-and-desist letter, Downhill Battle set up the corollary site GreyTuesday.org and started organizing "Grey Tuesday" as a day that activists would distribute *The Grey Album* over the Internet in violation of copyright law. In its announcement, Downhill Battle stated:

> This first-of-its-kind protest signals a refusal to let major label lawyers control what musicians can create and what the public can hear. *The Grey Album* is only one of the thousands of legitimate and valuable efforts that have been stifled by the record industry—not to mention the ones that were never even attempted because of the current legal climate. We cannot allow these corporations to continue censoring art; we need common-sense reforms to copyright law that can make sampling legal and practical for artists. (Downhill Battle 2004)

When Downhill Battle announced its plans, it received an overwhelming wave of support from individuals and Web site owners, who volunteered to distribute the album. Downhill Battle reported on its site that "more than 400 sites are currently listed as participating in the protest. At this point, we are receiving hundreds of emails from sites that are going grey and we simply can't keep that list up to date, but that support means a lot and we are thrilled that so many sites are getting involved" (Downhill Battle 2004).

EMI had sent its cease-and-desist letter on February 14, 2004, and on Tuesday, February 24, 2004, activists around the world hosted *The Grey Album* for twenty-four hours. At the end of the day, Downhill reported: "After a survey of the sites that hosted files during Grey Tuesday, and an analysis of filesharing activity on that day, we can confidently report that the Grey Album was the number one album in the US on February 24 by a large margin. Danger Mouse moved more 'units' than Norah Jones and Kanye West, with well over 100,000 copies downloaded. That's more than 1 million digital tracks" (Downhill Battle 2005a).

Although these figures have not been independently verified, the *New York Times* reported that "Greytuesday.org reached the top ranking on Blogdex and Popdex, Web sites that track which sites are being linked to from blogs" (Werde 2004). Thus, although the figures reported by Downhill are difficult to validate, the fact that its site was linked so extensively and that *The Grey Album* remains available on peer-to-peer networks to this day, suggests that if the numbers were not achieved in February 24, 2004, they certainly must have been by now.

Besides being critically lauded by *Rolling Stone*, the *Boston Globe*, *The New Yorker*, and other outlets, media coverage of *The Grey Album* and Grey Tuesday was widespread and contributed to the event's reported success. Burton was portrayed as an artist doing art for art's sake. When asked by *The New Yorker* if he thought the album would ever be released commercially, he noted, "Hell, no. . . . That's one of the things I struggled with. I told myself, 'Never will this come out. . . . Must still do . . . must still do" (quoted in Greenman 2004). The media's response to Grey Tuesday was largely positive. *MTV News, AllHipHop News, Wired News,* the BBC, *Billboard Magazine,* the *Boston Globe,* Reuters, and other news outlets reported the story, portraying Burton as an avant-garde artist experimenting with a new and exciting medium.

In response to Grey Tuesday, EMI sent more than 150 cease-and-desist letters to the Web site owners it could identify, only to be largely ignored. By all accounts, the Grey Tuesday event was a great success for Downhill Battle and for the digital rights movement. It reached a wide constituency and made record companies seem like bullies for using copyright law to stifle creativity and intimidate music listeners.

Grey Tuesday spawned a new project at Downhill Battle. The Web site BannedMusic.org was meant to showcase work similar to Burton's, all with a clear political message regarding copyright and sampling. For example, one of BannedMusic's projects was an exhibit called Illegal Art, and on that section of the site one could obtain a CD of sample-heavy music (itself titled *Illegal Art*), whose liner notes make clear the distributors' commitments regarding copyright and sampling:

> For our culture to be a space for free expression and for creativity to flourish, audio artists must be able to build on bits and pieces of preexisting music. While the "fair use" doctrine allows artists to appropriate other works, it does so only in cases of commentary or parody. Fair use doesn't apply to the majority of "second-takers," those artists who reuse sounds without directly referring to the original. Most of these tracks would never have existed if the artists had adhered to copyright law. Many other works might never be heard unless we act soon to grant artists the right to create them. ("Illegal Art Audio" 2004)

Creating digital outlets for music that violates copyright law as an act of civil disobedience was not the only tactic in Downhill's toolkit. It also sponsored a section on its site called "Battle Labs."

Downhill Battle's Battle Labs: Restructuring Technological Enforcement

Battle Labs was designed to coordinate and introduce software consistent with Downhill Battle's mission. It made "free, open-source software for online organizing and strategic file-sharing" (Downhill Battle 2005b).

Battle Labs[3] worked on or proposed a number of projects. Of the projects listed in table 8.5, all are open-source projects collaboratively developed on the SourceForge.net site. They fall into two categories: resource-management technologies and resistance technologies. The resource-management technologies, such as Battle Cart, Defense Fund, and Donation Bats, have helped organizations manage fund-raising. Local Link, Tabling

Table 8.5
Technology-Development Projects for Downhill Battle's Battle Labs Circa 2005

Technology	Function	Status of Project
Local Ink	Server-side application that will search for newspapers in user's area and generate a form "Letter to the Editor" regarding restrictive copyright issue. Also allows the user to write letter.	Developed
Battle Cart	Server-side application that allows small organizations such as nonprofits to sell items for fund-raising purposes. It interfaces with PayPal, a trusted online bill-paying service.	Developed
BlogTorrent	Allows users to upload movies and share them on their blogs using the BitTorrent file-sharing protocol.	Developed
Broadcast Machine	Next-generation BlogTorrent allows users to set up RSS feeds for video content on their blogs. Thus, a user can establish channels similar to RSS text channels. Using the BitTorrent file-sharing protocol, transfers are theoretically faster and bandwidth shared among blog readers who have the torrent seed for the video file.	Developed
Defense Fund	Server-side application that allows activists to collect donations for multiple entities. Donations are routed through PayPal and distributed evenly among the group automatically.	Developed
iTMS4ALL	"Perl script that can access Apple's iTunes Music Store."	Not functional
DHB Chapter Sites	Web site development kit to help activists easily set up Web sites.	Under development
Copy Finder	Networking tool that helped activists in a network list resources available to them such as copy services, supplies, or other essentials. The application allowed users to find each other based on these needs.	Under development
Tabling Database	"The software will keep a calendar database of who's doing what on what day, and automatically sends email reminders beforehand and a request for feedback afterwards. This would also function as a general calendar of events where there would be a Downhill Battle presence."	Developed
Donation Bats	Mirroring Howard Dean's donation tracker, a tool that will integrate with Battle Cart and graphically track progress toward a fund-raising goal.	Under development
Local Wi-Fi File Sharing	Easy-to-use application that will allow a user to use mobile computers to set up a wireless peer-to-peer server.	Developed but needs to be made user-friendly

Source: From Downhill Battle 2005b.

Database, and DHB Chapter Sites helped Downhill Battle manage its people by streamlining scheduling for pamphletting, Web site design, and letter writing, respectively. Resistance technologies do not manage resources but rather are subversive by creating technological architectures that can help the digital rights movement live out its mission. Local Wi-Fi File Sharing, iTMS4All, Broadcast Machine, and BlogTorrent are examples of resistance technologies designed by Downhill Battle. Local Wi-Fi File Sharing untethers peer-to-peer file sharing from the confines of fixed servers; iTMS4All can undo DRM for iTunes, and Broadcast Machine and BlogTorrent allow users to turn their blogs into video broadcast stations using the BitTorrent protocol, which greatly improves file-transfer efficiency.

Broadcast Machine and BlogTorrent in particular are technologies of note. They bring together important social-computing applications that have revolutionized participatory culture online. They harness the power of a blog's streamlined content production and distribution system by adding a layer of efficiency and ease to the blog's ability to distribute more complex content (such as video) via Really Simple Syndication (RSS) feeds. They do this by melding the blog technology with BitTorrent, a file-sharing protocol that distributes the bandwidth required to transfer large files over the populations in the network. With each host contributing a piece of the file to the downloader, the bandwidth usage is shared by the community as a whole, and the file is transferred quickly. The person downloading from the network also subsequently becomes a server, distributing the bits he or she has downloaded to others who need them even though her own download is not yet complete. Therefore, although many on the network may start off with incomplete versions of a desired file, so long as there is at least one full copy on the network and the network is up, all will eventually have the file and will get it faster than if it were transferred from a single complete source host. The network, then, is a "torrent of bits" that together make the whole file. At the risk of sounding overly optimistic, I believe that Broadcast Machine and BlogTorrent are truly revolutionary because they allow the lone user to have the power to broadcast video over the net efficiently—something that only major outlets that can afford the bandwidth can currently do. If current trends in social computing and participatory culture on the Internet continue, then these types of technologies will greatly help the process of making broadcasters out of all who have access to a computer and an Internet connection.

It is important here to make a further distinction among the examples of resistance technologies discussed throughout the book. Some of those technologies are infrastructures of distribution, and so although not directly circumventing copy-protection measures, they serve to create the technological world where the practice of sharing content is possible. Other technologies such as iTMS4All were circumvention technologies, and those resistance technologies serve to gain access to content. Together, infrastructure and circumvention tools form a repertoire of technological affordances that can realize movement goals.

The Battle Labs' work is the best example of an organized attempt to use technological resistance within the context of an SMO. In fact, Downhill Battle was the only organization that did this at the time I conducted my research from 2005 to 2006. All other technologies that one might consider resistance technologies have come from hacker communities that cannot be classified as SMOs in any conventional sense.

This brings us to a fourth key player in the movement—a player who does not fit the model of an SMO and is not mentioned in my introduction to this chapter, but who has had an important impact on the movement: the hacker.

The Hacker as a Force in the Movement

I discussed earlier the work of hackers such as Jon Johansen and the developers of Hymn and other iTunes hacks, illustrating how hackers' rapid responses and continuous development put constant pressure on media outlets such as the iTMS to continue to devise ways of technologically safeguarding content and contracts. The same can be said of other instances where technological protection mechanisms were challenged by circumvention technologies designed to expand user access to copyrighted content, as in CSS and DeCSS, eBook protection and the AEBPR, and FairPlay and the iTunes Hacks. The presence of skilled activists and hackers who continue to design technological resistance has important consequences for the social movement structure, specifically with regard to how direct action is applied against the establishment.

In the conventional sense, direct action involves disturbances of the ordered practices of the system under attack. These disturbances include sit-ins, marches that block streets, and violent direct actions such as clashes with police or other authorities. In the case of using technological means as a form of direct action, scholarship on what Tim Jordan (2002) has called "hacktivism" points to an emerging use of technological disturbances that can be characterized as mass virtual direct action (MVDA). In MVDA, large numbers of activists employ information communication technology such as the Internet to engage in disruptive information politics, where the flows and ownership of information are challenged via technological means. For example, denial-of-service attacks are made on corporation sites by either coordinating many activists to continuously click on a site or designing an application that does this automatically. In either case, the target site is incapacitated by the overload of requests. These types of direct action can be performed by many activists or by a single hacker.

Technological resistance goes further than MVDA by creating alternative technologies whose adoption and continued use structure behaviors consistent with the goals of the digital rights movement. Thus, iTunes hacks such as PyMusique are important not only because they defeat the iTunes DRM system, but because they create a technological world without DRM by replacing iTunes altogether. Technologies that create distribution architecture such as peer-to-peer file sharing or those highlighted in the case of Battle Labs also do this.

Furthermore, adoption of technological resistance restructures the relationship between SMOs and collective action. Collective action is traditionally dependent on some level of centralized organization that can coordinate participants for the collective-action event—a sit-in or a march, for example (McAdam, McCarthy, and Zald 1996). With technological resistance, the hackers or the designers of these technologies become important, and the use of the technology by many becomes the collective action. Given that the Internet can help distribute technological resistance technologies to many participants, one lone hacker or hacker community can be the driving force behind collective use of a technology of resistance. No longer are the SMO's resources needed as much for this type of coordination, and the level of ideological commitment from participants can be slight because they may simply have functional commitment to the technology as opposed to believing in the politics of the technology.

This finding reflects discussion within both social movement studies and network studies on the possibility of effecting change and organized collective action without organizations, the impact of the Internet on traditional means of organization, and the impact of social networks on organizational importance in some contexts (Earl and Schussman 2003; Postigo 2010; Shirky 2008; Tarrow 1994).

Conclusion

Analysis of the digital rights movement's tactical repertoire illustrates the use of information communication technology and software in acts of "hacktivism" to further a political agenda. Importantly, the movement seeks to alter the status quo in digital copyright by undermining the established order by committing acts of disobedience and by creating alternative legal and technological worlds. The work of Creative Commons to create alternative licensing and of hackers to design applications that circumvent or make DRM impossible illustrates this trend. Although SMOs play an important role in the digital rights movement (the EFF being a key organization), use of technological resistance has placed the hacker or anyone technologically savvy enough to design technological resistance in a position of prominence in the movement.

It is important to note that the names of individual hackers seldom come to mind when one considers the network of SMOs currently active in the movement (as seen in table 8.5). Hacker activities are typically outside the realm of what SMOs do. If SMOs were to engage in hacking, it would undermine their credibility in institutional environments. Thus, the EFF does not link to any hackers creating anti-DRM technology, but it does encourage them indirectly (in 2001 the EFF awarded Jon Johansen an Innovation Award for his work on DeCSS). Analysis of the movement's tactics shows a division of labor, where most SMOs stay an institutional course to maintain legitimacy, and hackers work behind the scenes to achieve technological resistance. Only Downhill Battle actively engaged in the design of technologies meant to undermine technological copyright protection. The division of labor can be attributed to institutional histories. SMOs do what they do best: they mobilize the resources they are most familiar with. Historically, for example, the EFF has been involved in courtroom battles and in the halls of Congress. It is an organization that needs to retain legal legitimacy if it wishes to continue its fight in these arenas, so its design and distribution of illegal technologies would greatly affect that legitimacy.

Hackers play an important role, too, though, because they do not formally belong to the SMOs, yet use of their technologies has the disruptive effects of collective action. This fact gives the movement a novel dynamic that places much power in the hands of a few savvy technologists.

9 Conclusion

The legislative history of the DMCA shows the blind spots in its configurations of users and their activities. The technological realities of the NII raised some troublesome issues for lawmakers and media companies, putting pressure on longstanding applications of copyright law. Awareness of the blind spots in the legislative process instigated the work of the digital rights movement and its progress toward challenging technolegal regimes over content so that consumers can do more with their legally purchased media.

The formulation of the DMCA, informed by the IITF WGIP, was a contentious process where stakeholders from the content industry, libraries, and the technology industry sought to carve out exemptions and expansions in copyright. Yet during this process very few organizations represented consumers. The policy proposal that emerged from the WGIP hearings was informed by fear of what would happen to intellectual property if law and technology did not protect it on the information superhighway. This generalized fear led to restrictive technology policies that, instead of harnessing the power of digital technology for the user, harnessed technology for copyright owners, allowing them to design technologies that would enforce ever more restrictive licenses and constrict traditional privileges such as first sale and fair use.

In response to these developments, a core group of activists (technologists and law professors at first) called for better policy that would allow consumers the privileges that they enjoyed in analog media. Furthermore, these activists wanted to expand those privileges so that users could become active creators. As a response to the formulation of the DMCA, the digital rights movement began advocating for fair copyright law. The movement's advocacy was founded on the ideals of legal scholars such as James Boyle (1997), who thought that consumption of cultural products ought to be politicized and that consumers ought to feel their stake in culture every time they listen to music, read a book, or view a film.

The technolegal structures that emerged from the policy wrangling gave rise to a host of cases that challenged the law on a number of constitutional and technological grounds. These cases illustrate the issues that copyright protection in digital media would create. Much like policymakers who struggled with the changes that legally defined categories such as phonorecords and transmissions would undergo given shifting technological realities in the early twentieth century, courts and activists in the late twentieth and early twenty-first centuries were left to contend with the changes that law and technology would bring to the way users saw fair use, free speech, and their own personal relationship with content. In response to those cases and threats to their perceived rights, activists formulated a host of strategies from lawsuits to hacks. Although the legal or institutional strategies were important, it was the technological forms of resistance that distinguished digital rights activism. The role of technology in the movement demonstrates how activism might take shape for other causes that deal intimately with the technological.

For the digital rights movement, technology is a resource for a number of reasons. From a basic point of view, it is a resource in the sense that access to it can mean the difference between being part of cultural production or not. Computers, the Internet, DVD players, and other devices and processes that allow users to experience content are thus important at a basic level. But technology is also a resource for activism. The very same technologies used to consume can morph, be hacked, and worked around to allow for specific participatory uses or conveniences. Technology, then, is a resource not only to experience content, but to modify it and to participate in it. However, technology is also a resource for the content industry because it mediates that industry's goods and serves to regulate their consumption. It is a resource for both sides of the digital rights debate for not dissimilar reasons: it mediates delivery, participation, and consumption of content in opposite ways depending on what side of the debate one finds oneself. The fight between the content industry and digital rights activists, then, is in part a fight over this resource, where both try to define its meaning and its uses through legal strategies, licensing mechanisms, and architectures of consumption or participation.

The meaning of technology as a resource cannot be underestimated here. Whether one uses Mozilla or Explorer or Safari; whether one uses Winamp, VLC, or iTunes; whether one uses Unbutu, Windows, or Mac OS, all these uses open to framing that goes beyond the actual functionality of those technologies. For many, technological use is a political brand, and what one uses signals where on a particular political debate one might lie.

Thus, thinking of technology as a resource is to think of it not only as a functional resource, but also as a linguistic resource. Technology speaks a certain kind of language about cultural production—a participatory ethos and its politics.

Technology can also be thought of as opportunity in the digital rights movement—an opportunity to allow for a particular relationship between consumers and content. Without the current technological realties, the idea of a culture of participation on the scale imagined by the movement would not be possible. Likewise, for the content industry, without digital technologies the idea of enforceable licensing schemes that regulate user consumption to a high degree and that increasingly sanction participation in production of content using existing intellectual properties would not be possible. In the context of the struggle between the movement and the industry, the regimes of digital technology can serve as a form of opportunity to rewrite the architectures of consumption into architectures of participation.

The idea of technology as opportunity is not divorced from its role as resource. In the sense that technology can serve as a meaningful resource, those meanings imparted on the technology can give the movement and its activists the language to talk about users' lived consumptive and productive experiences. They make possible the opportunity to talk about cultural participation where perhaps that opportunity did not exist previously. This is why the case studies presented in this book involve not just technologies, but the legal arguments about the meaning of fair use and free speech that those technologies made coherent. Arguments about what constitutes free speech and fair use are framed in language that implicitly (and explicitly) references the technological realities of technology users' lives.

The central role of technology is not deterministic, and this analysis should not be read as such. If anything, this text recognizes many determinisms. There is a very real social construction in the formulation of the meaning of technologies such as iTunes hacks or the products of Battle Labs. But we should not shy away from pointing out the clear ways in which technology, whether designed by movement activists or the content industry, attempts to structure not only the user's experience with cultural goods, but also his or her views on that experience. Technologies of this sort are in many ways normative moves executed in attempts to convince users that the world ought to be a certain way. Technology, then, is a space of contest where the players in this particular struggle come to realize their worldviews and convince (or cajole) others into embracing those views.

What is the meaning of the digital rights movement, and what does it show us about technology for society as a whole? What it shows primarily is that as various forms of consumption are increasingly mediated through technologies that can increasingly control our levels of access and involvement, it becomes important to seize that very same technology for the opportunities it may afford us to become participants in the making of cultural goods. This capture requires not a tacit acceptance of the means provided for us by media companies, but rather a consideration of how we might actively design technologies for ourselves. In that sense, media consumption should be a form of intervention into the manufacture of cultural goods, and the technologies we choose to mediate content should have those affordances. Two decades ago Langdon Winner (1985) wrote about the politics of artifacts when thinking about the way technologies have the potential both to structure social configurations and to influence political arrangements (supporting or undermining democratic principles). If technologies have these potentials, then technological use is an implicit political exercise, for it is only when we willingly adopt technological systems that they become embedded in the social architecture and gain their formidable power to influence society in certain ideological ways.

The movement for digital rights has shown this to be true, and its politics—its technological political reality—explicitly reflects this view. Today, as I mentioned in the introduction to this book, the movement has expanded to engage a wider discourse of cultural rights, arguing for a citizenry's right to engage and reconfigure the cultural forms generated by the media industries. Many of the actors covered in this text have changed since the early days of activism. Lawrence Lessig, for example, now works on other issues even as his ideas remain powerful in the movement's underlying logic. Although the movement continues to be intimately tied to the technological, the term *digital* to describe the types of rights it seeks seems limiting, and *participatory rights* is perhaps more apt. In addition, some of the organizations and key players illustrated in some of the case studies have changed. Downhill Battle, for example, is no longer active, but its founders went on to start the Participatory Culture Foundation, which is responsible for producing Miro, an open-source video player meant to decenter concentration of user-generated video within any one online platform (YouTube, for example), and Jon Johansen is now working in Silicone Valley.

Despite these changes, the movement continues as other players and issues take the field. Corporate actors are now also entering the field, for example, as the interests of Internet giants like Google, banking on the continued creativity and participation of users, overlap with those of the digital rights movement. Increasingly they will join the protests against copyright protection laws seen as overreaching and that threaten their business models. Most recently, this was exemplified in Google's participation in the successful protest against the Stop Online Piracy Act (SOPA), where the Web stalwart blacked out its famous logo and pointed users to means of contacting their legislative representatives.

The movement also now seems more global, with activists in Europe taking the lead in articulating the flows of a global technolegal regime that seeks to circumscribe the public domain. Student groups are also more active in this regard, with the free-culture student movement in the United States and groups such as Isaac Hackimov and the Hackademy in Spain. Using cultural jamming and hacking to achieve their goals of raising awareness, the young (our digital natives) now seem to be the most active originators of technological resistance and interventions.

This text, then, should be read as an opening exploration of a movement that today is global and broad in scope. I hope it is a good beginning to an ongoing study of this issue, one that will continue through my own future work, but also that of others.

Appendix

Table A.1
Organizations and Individuals Submitting Comments Prior to the NII WGIP Public Hearings in 1994

Representative	Organization
Steven J. Metalitz	Information Industry Association
Maria Pallante	National Writers Union
Stephen Haynes	West Publishing
Lisa Freeman	Association of American University Presses
Robert Oakley	on behalf of several library and educational associations
Joseph Cosgrove	no organization
Denise Bybee	International Society for Technology in Education
David Rothman	no organization
Arnold Lutzker, Michael Goldstein, David Pierce, and Richard Marks	American Association of Community Colleges
Fritz Attaway	MPAA and RIAA
Richard Ducey	National Association of Broadcasters
Edward Murphy	National Music Publishers Association
John Masten	New York Public Library
Fary Griswold	Infologic Software
Robert Kahn	Corporation for National Research Initiatives
Brad Cox	Center for Electronic Markets
Thomas Lemberg	Business Software Alliance and Alliance to Promote Software Innovation
Ronald Laurie	no organization
Ronald Palenske	Information Technology Association of America
Mark Traphagen	Association of Software Publishers
Brian Kahin	Interactive Multimedia Association

Table A.1
(continued)

Representative	Organization
Gary Shapiro	Consumer Electronics Group of the Electronic Industries Association and the Home Recording Rights Coalition
Douglas Brotz	Adobe Systems Inc.
Frank Connolly	no organization
Nicholas Veliotes	Association of American Publishers
Andrew Oram	no organization
Greg Buell	no organization
Albert Teich	American Association for the Advancement of Science
Joseph Alen	Copyright Clearance Center, Inc.
Albert Henderson	no organization
Morton Gould and Fred Koenigsberg	American Society of Composers, Authors, and Publishers
Timothy King	John Wiley and Sons
Walter Biggs	no organization
Gregory Ahoronian	Source Translation and Optimization
David Roland	Roland Projects
Chad Huston	Schlumberger Laboratory for Computer Science
Simon Higgs	no organization
Bernard Sorkin	Time Warner
Jo Clare Peterman	no organization
Thomas Galvin	no organization
Alan Hodson	no organization
Martin Weiss	no organization
Henry H. Perritt Jr.	Villanova University
Kerric Harvey	no organization
Chuck Kolbenson	Summa Four
George Bynon	University of California at Davis Library
Cornelius Pings	Association of American Universities
Edward Valauskas	American Library Association
Benjamin Ivins	National Association of Broadcasters
Brian Kahin	Information Infrastructure Project
Gregory Ferenbach and Paula Jameson	PBS
Peter Choy	American Committee for Interoperable Systems
Carol Gottlieb, Arnold Lutzker, Martin Scorsese, Elliot Silverstein, and Robert Wise	Writers Guild of America, Artists Rights Foundation, Directors Guild of America

Table A.1
(continued)

Representative	Organization
Daniel Brenner and Diane Burnstein	National Cable Television Association, Inc.
Lance Rose	Association of Shareware Professionals
Daniel Abraham	Graphic Artists Guild
committee members	Federal Networking Council Advisory Committee
Ronald Myrick	Intellectual Property Law Section of the American Bar Association
Rhett Dawson	Computer and Business Equipment Manufacturers Association
R. S. Talab	no organization
Caron Hughes	Research Libraries Group Inc.
Christopher Hyun	Arts Management International
Michael Goldstein	Distance Learning Institutions
Christopher Hyun	New York County Lawyers Association and Committee on Communications and Entertainment Law
Andre Paul	Satellite Broadcasting and Communication Association
Theodor Nelson	Xanadu On-line Publishing
Joseph Clark	Video Discovery
Thomas Lee	no organization

Table A.2
Organizations and Individuals Submitting Comments after the Release of the Green Paper and for the IITF WGIP Public Hearings in Washington, DC, Los Angeles, and Chicago, 1994

Witness	Organization
Daniel Abraham	Graphic Arts Guild
Geoffrey Adams	British Copyright
Paul Aiken	Authors' League
Joseph Alen	Copyright Clearance Center
Allen Arlow	Computer and Communications Industry Association
Diane Balestri	Princeton University
Chris Barlas	Working Group on Copyright and Technology, British Copyright Council
William Barlow and Robert Steinberg	Times Mirror Company
Alan Batie	no organization

Table A.2
(continued)

Witness	Organization
Henry Baumann and Benjamin Ivins	National Association of Broadcasters
David Bender	Special Libraries Association
Marvin Berenon	BMI
Marilyn Bergman and Fred Koenigsberg	American Society of Composers, Authors, and Publishers
Geoffrey Berkin	no organization
Joe Jekovitz	Houghton Mifflin
John Berry	University of Illinois
Jame Bikoff	Nintendo of America
Carol Billings	American Association of Law Libraries
Kathleen Bloomberg and Jane Running	Illinois State Library
Cynthia Braddon	McGraw-Hill
Lorin Brennan	American Film Marketing Association
Steven Ames Brown	Artists' Rights
Thomas Bonetti	Celebrity Licensing Inc.
Scott Busby	
Kaye Cladwell	Software Industry Coalition
Alan Carey	Picture Agency Council of America
Peter Choy	American Committee for Interoperable Systems
Kenneth Crews	Indiana University Law School
Jeffrey Cunard	America Online, Compuserve, Delphi Internet Services, GE Information Services, Lexis Counsel Connect, Prodigy Services, Lance Rose and Associates, Ziff Communicaitons
Arthur Curley	American Library Association
David Curtis	Microsoft for Business Software Alliance and Alliance to Promote Software Innovation
Willam Daniels	Paul and Stuart
James Davis	Xerox
Rhett Dawson, Robert Holleyman, and Emery Simon	Computer Business Equipment Manufacturers Association, Business Software Alliance, Alliance to Promote Software Innovation
Donna Demac	Institute for Learning Technology, Columbia University
Sarah Deutch	Bell Atlantic
John Dill	Mosby–Year Book
William Ellis	IBM
Gregory Ferenbach	PBS
Carl Fornaris and Robert Garrett	Submitted on behalf of the Office of the Commissioner of Baseball

Table A.2
(continued)

Witness	Organization
Roy Freed	no organization
David Friedman	University of Chicago Law School
Laura Gassaway	University of North Carolina Law Library
Branko Gerovac and Richard Solomon	MIT
Jane Ginsburg	Columbia University
Professor Mitchell Golden	
George Gross	Magazine Publishers of America
Czeslaw Grycz	University of California at Los Angeles
David Guttman	no organization
Colin Hadley	Copyright Licensing Agency
Trotter Hardy	Marshall-Whythe School of Law, College of William and Mary
Ann Harkins, Joe Waz, and Michele Woodward	Creative Incentive Coalition
Bruce Hayden	no organization
R. H. Hedgzi	no organization
Professor Lee Holloar	no organization
Linda Hopkins	Intelliware
Linda Hopkins	Subcommittee on Copyrights of the American Bar Association and the NII
Irving Horowitz	Transaction Publishers (also a professor at Rutgers University)
John Howard	no organization
James Claudia	Committee for America's Copyright Community
Mary Brandt Jensen	University of South Dakota
Richard Johnson	River of Stars Software Development
Michael Joyce	Vassar College
Julia Kane and Martin Taschdjian	US West Inc.
Mahatma Kane-Jeeves	no organization
Menelaos Karamichalis	Mallinckrodt Institute of Radiology
Abraham Katz	United States Council for International Business
Kenneth Kaufman	SESAC Inc.
John Kelly	Recording for the Blind
Charles Kerns	Stanford University
Jack King	Coalition for Consumers' Picture Rights
Leila Kinney	College of Arts Association (CAA): Committee on Electronic Information
Donald Kiser	Grain Processing Corporations
Susan Kornfield	Bodman, Longley and Dahling
Ellen Kozak	Niles and Niles

Table A.2
(continued)

Witness	Organization
Al Lauck	no organization
David Leibowitz	RIAA
Mark Lemley and Neil Natanel	University of Texas School of Law
Susan Lesch	AOL
Howard Liberman	Primosphere Limited Partnership
Jessica Litman	Wayne State University
Lydia Pallas Loren	Bodman, Longley and Dahling
Nicholas Lowe	Performing Rights Society for Music, London
Arnold Lutzker	Artists' Rights Foundation
Stuart Lynn	Commission on Preservation and Access
Michael Malone	Gryphon Software
Joe Mambertti	University of Chicago
Edward Massie	CCH Inc.
Gottfried Mayer-Kress	Center for Complex Systems Research, Beckman Institute
Philip McAleer	Maineville Products
Steven J. Metalitz	Information Industry Association
Theodore Miles	National Public Library
David Moran	Dow Jones and Company
Lynn Morgan	Association of Academic Health Sciences Library Directories and Medical Library Association
Edward Murphy	National Music Publishers Association
John Ogilvie	Madison and Metcalf
Charles Ossolla	American Society of Media Publishers
Michael J. Pierce and Kenneth Salomon	Dow, Lohnes and Albertson for a number of higher-education institutions
Mary Beth Peters	US Registrar of Copyrights
Marshall Phelps	IBM
Billy Barron Plano	
James Popham	Association of Independent TV Stations
Anssi Porttikivi	
F. E. Potts	ACS Publishing
Dr. Bojan Pretnar	Industrial Property Protection Office, Slovenia
John Rademacher	American Farm Bureau Association
Anita Rivas	Artists Manager
Pat Rogers	Nashville Songwriters Association International
Lance Rose	Association of Shareware Professionals
Victor Rosenberg	Personal Bibliographic Software, Inc.
Cynthia Russo	National Council of Teachers of Mathematics

Table A.2
(continued)

Witness	Organization
David Rothman	no organization
Richard Koman	Individual Consumer Rights, O'Reilly Publishers
Arthur Rubin	no organization
John-Willy Rudolph	Kopinor–the Reproduction Rights Organization of Norway
William Ryan	AT&T
Arthur Sackler	Time Warner
Pamela Samuelson	University of Pittsburg
James Schatz	West Publishing
Gary Shapiro	Home Recording Rights Coalition
David Shirley	Pennsylvania State University
Dick Shoemaker	National PC Users Group
Victor Siber	IBM
Robert Simons	International Intellectual Property Alliance
Bill Sohl	no organization
Janet Staiger	Society for Cinema Studies
Randall Stempler	Infosafe Systems
August Steinhilber	National School Boards Association
John Sturm	Newspaper Association of America
Christine Sundt	no organization
John Sutton	Heller Ehrman White and McAuliffe
Janice Tanne	American Society of Journalists
Walter Thompson	Vanderbilt University
Mark Traphagen	Software Publishers Association
Scott Turow	Authors' League
John Vaughn	Association of American Universities
Edward J. Valauskas	American Library Association
Nicholas Veliotes	Association of American Publishers
Wim Vestappen	Vevam (Netherlands)
Walt Wahnsiedler	no organization
Priscilla Walter	Gardner, Carter and Douglas
Sandra Walker	Visual Resources Association
Ginger Warbis	unknown
Daniel Warren	Newsletter Publishers Association
Duane Webster	Association of Research Libraries
Gloria Werner	Association of Research Libraries
Sarah Wiant	Washington and Lee School of Law library
Joshua Yeidel	Learning Systems
Ronald Yin	Limbach and Limbach
Toyomaro Yoshida	Institute of Intellectual Property

Table A.3
Digital Rights Movement Organizations, Mission Statements, and Classification

Organization	Mission Statement	Class
Creative Commons	"To build a layer of reasonable, flexible copyright in the face of increasingly restrictive default rules" (Creative Commons 2005).	nongovernmental organization (NGO)
Free Software Foundation (FSF)	"Dedicated to promoting computer users' rights to use, study, copy, modify, and redistribute computer programs. The FSF promotes the development and use of free software, particularly the GNU operating system, used widely in its GNU/Linux variant" (FSF 2005).	NGO
Samuelson Law, Technology, and Public Policy Clinic at the Berkeley Boalt School of Law	"The clinic aims to serve as the public's voice in legal and regulatory disputes presently dominated by lobbyists and the government. The Clinic takes on projects in many fields relating to the public interest in technology. Areas we are currently focusing on include: Copyright, Digital Rights Management, Free Speech, Open Source and Privacy" (Samuelson Law, Technology, and Public Policy Clinic 2005).	law school
Electronic Frontier Foundation (EFF)	"When our freedoms in the networked world come under attack, the Electronic Frontier Foundation (EFF) is the first line of defense. EFF broke new ground when it was founded in 1990—well before the Internet was on most people's radar—and continues to confront cutting-edge issues defending free speech, privacy, innovation, and consumer rights today. From the beginning, EFF has championed the public interest in every critical battle affecting digital rights. Blending the expertise of lawyers, policy analysts, activists, and technologists, EFF achieves significant victories on behalf of consumers and the general public. EFF fights for freedom primarily in the courts, bringing and defending lawsuits even when that means taking on the US government or large corporations. By mobilizing more than 50,000 concerned citizens through our Action Center, EFF beats back bad legislation. In addition to advising policymakers, EFF educates the press and public. Sometimes just defending technologies isn't enough, so EFF also supports the development of freedom-enhancing inventions" (EFF 2005).	NGO

Table A.3
(continued)

Organization	Mission Statement	Class
Global Internet Liberty Campaign (GILC)	"Advocates: Prohibiting prior censorship of on-line communication; Requiring that laws restricting the content of on-line speech distinguish between the liability of content providers and the liability of data carriers; Insisting that on-line free expression not be restricted by indirect means such as excessively restrictive governmental or private controls over computer hardware or software, telecommunications infrastructure, or other essential components of the Internet; Including citizens in the Global Information Infrastructure (GII) development process from countries that are currently unstable economically, have insufficient infrastructure, or lack sophisticated technology; Prohibiting discrimination on the basis of race, color, sex, language, religion, political or other opinion, national or social origin, property, birth or other status; Ensuring that personal information generated on the GII for one purpose is not used for an unrelated purpose or disclosed without the person's informed consent and enabling individuals to review personal information on the internet [*sic*] and to correct inaccurate information" (GILC 2005).	NGO
Lawrence Lessig Blog	Blog that reports on news concerning digital copyright and the work of Creative Commons.	blog
Electronic Privacy Information Center (EPIC)	"EPIC was established in 1994 to focus public attention on emerging civil liberties issues and to protect privacy, the First Amendment, and constitutional values" (EPIC 2005).	NGO
Copyfight	"Explore[s] the nexus of legal rulings, Capitol Hill policy-making, technical standards development, and technological innovation that creates—and will recreate—the networked world as we know it. Among the topics we'll touch on: intellectual property conflicts, technical architecture and innovation, the evolution of copyright, private vs. public interests in Net policy-making, lobbying and the law, and more" (Copyfight 2005).	blog

Table A.3
(continued)

Organization	Mission Statement	Class
Berkman Center for Internet and Society at the Harvard Law School	Dedicated to providing assistance in digital copyright cases.	law school
Center for Democracy and Technology (CDT)	"The Center for Democracy and Technology works to promote democratic values and constitutional liberties in the digital age. With expertise in law, technology, and policy, CDT seeks practical solutions to enhance free expression and privacy in global communications technologies. CDT is dedicated to building consensus among all parties interested in the future of the Internet and other new communications media" (CDT 2005).	NGO
Public Knowledge	"Public Knowledge is a group of lawyers, technologists, lobbyists, academics, volunteers and activists dedicated to fortifying and defending a vibrant information commons. Our first priority is to stop any bad legislation from passing—laws we think would slow technology innovation, pick market winners, shrink the public domain, or prevent fair use" (Public Knowledge 2005).	NGO
Center for Internet and Society (CIS) at the Stanford Law School	"The Center for Internet and Society (CIS) is a public interest technology law and policy program at Stanford Law School and a part of [the] Law, Science and Technology Program at Stanford Law School. The CIS brings together scholars, academics, legislators, students, programmers, security researchers, and scientists to study the interaction of new technologies and the law and to examine how the synergy between the two can either promote or harm public goods like free speech, privacy, public commons, diversity, and scientific inquiry. The CIS strives as well to improve both technology and law, encouraging decision makers to design both as a means to further democratic values" (CIS 2005).	NGO

Table A.3
(continued)

Organization	Mission Statement	Class
Chilling Effects: Cease-and-Desist Clearinghouse	"A joint project of the Electronic Frontier Foundation and Harvard, Stanford, Berkeley, University of San Francisco, University of Maine, George Washington School of Law, and Santa Clara University School of Law clinics. Chilling Effects aims to help understand the protections that the First Amendment and intellectual property laws give to online activities. We are excited about the new opportunities the Internet offers individuals to express their views, parody politicians, celebrate their favorite movie stars, or criticize businesses. But we've noticed that not everyone feels the same way. Anecdotal evidence suggests that some individuals and corporations are using intellectual property and other laws to silence other online users. Chilling Effects encourages respect for intellectual property law, while frowning on its misuse to 'chill' legitimate activity. The website offers background material and explanations of the law for people whose websites deal with topics such as Fan Fiction, Copyright [and so on]" (Chilling Effects 2005).	NGO
Computer Professionals for Social Responsibility (CPSR)	"CPSR is a global organization promoting the responsible use of computer technology. CPSR educates policymakers and the public on a wide range of issues. CPSR has incubated numerous projects such as the Public Sphere Project, EPIC (the Electronic Privacy Information Center), the 21st Century Project, the Civil Society Project, and the CFP (Computers, Freedom & Privacy) Conference. Originally founded by U.S. computer scientists, CPSR now has members in over 30 countries on six continents" (CPSR 2005).	NGO

Table A.3
(continued)

Organization	Mission Statement	Class
American Libraries Association	"The Digital Age presents new challenges to fundamental copyright doctrines that are legal cornerstones of library services. Libraries are leaders in trying to maintain a balance of power between copyright holders and users, in keeping with the fundamental principles outlined in the Constitution and carefully crafted over the past 200 years. Libraries are perceived as a voice for the public good and our participation is often sought in 'friend of the court' briefs in important intellectual property cases. Our involvement extends to the international copyright arena where we also follow the treaties to which the U.S. is a signatory and which could influence the development of copyright changes at home" (American Libraries Association 2005).	NGO
Downhill Battle	"Downhill Battle is a non-profit organization working to break the major label monopoly of the record industry and put control back in the hands of musicians and fans. Downhill Battle is a collaborative project and we work with musicians, music fans, artists, and designers around the world. There is a core group of people working full-time, based in Worcester, MA. We see an unprecedented opportunity to create a decentralized music business and a level playing field for independent musicians and labels. We're doing everything we can to make that happen. Software development—done strategically—is probably the most effective way to change culture in a positive direction right now. We're especially looking for Python help and Win32 and OS X specific help. Check out Downhill Battle Labs, Blog Torrent, and Participatory Culture Foundation" (Downhill Battle 2005a).	NGO

Table A.3
(continued)

Organization	Mission Statement	Class
Students for Free Culture	"FreeCulture.org is a diverse, non-partisan group of students and young people who are working to get their peers involved in the free culture movement. Launched in April 2004 at Swarthmore College, FreeCulture.org has helped establish student groups at colleges and universities across the United States. Today, FreeCulture.org chapters exist at nine colleges, from Maine to California, with many more getting started around the world. Named after the book *Free Culture* by Stanford University law professor Lawrence Lessig, FreeCulture.org is part of a growing movement, with roots in the free software / open source community, media activists, creative artists and writers, and civil libertarians. Groups with which FreeCulture.org has collaborated include Creative Commons, the Electronic Frontier Foundation, Public Knowledge, and Downhill Battle. FreeCulture.org has four major functions: 1. Creating and providing resources for our chapters and for the general public 2. Outreach to youth and students 3. Networking with other people, companies and organizations in the free culture movement 4. Issue advocacy on behalf of our members" (Students for Free Culture 2005).	grassroots organization in transition to NGO

Table A.3
(continued)

Organization	Mission Statement	Class
Privacy Rights Clearinghouse	"Privacy Rights Clearinghouse (PRC) is a nonprofit consumer organization with a two-part mission—consumer information and consumer advocacy. The PRC's goals are to: 1. Raise consumers' awareness of how technology affects personal privacy. 2. Empower consumers to take action to control their own personal information by providing practical tips on privacy protection. 3. Respond to specific privacy-related complaints from consumers, intercede on their behalf, and, when appropriate, refer them to the proper organizations for further assistance. 4. Document the nature of consumers' complaints and questions about privacy in reports, testimony, and speeches and make them available to policy makers, industry representatives, consumer advocates, and the media. 5. Advocate for consumers' privacy rights in local, state, and federal public policy proceedings, including legislative testimony, regulatory agency hearings, task forces, and study commissions as well as conferences and workshops" (PRC 2005).	NGO
Digital Future Coalition (DFC)	"Digital Future Coalition (DFC) is committed to striking an appropriate balance in law and public policy between protecting intellectual property and affording public access to it. The DFC is the result of a unique collaboration of many of the nation's leading non-profit educational, scholarly, library, and consumer groups, together with major commercial trade associations representing leaders in the consumer electronics, telecommunications, computer, and network access industries. Some key issues and proposals: Fair Use—Temporary Copies—First Sale—Preemption—Distance Learning—Library Exemptions—Anti-Circumvention and Copyright Management Information" (DFC 2005).	NGO

Table A.3
(continued)

Organization	Mission Statement	Class
Participatory Culture Foundation/Get Democracy	"Television is the defining medium of our culture. There's now an opportunity to create a television culture that is fluid, diverse, exciting, and beautiful. Built by people working together. * Get Democracy is developed by the Participatory Culture Foundation. * We're based in Worcester, Massachusetts. * We're a not-for-profit organization (501c3 pending). * We think it's a problem that a small number of corporations control mass media. * We think free, open-source, open standards internet [*sic*] TV is our best shot at a solution" (Participatory Culture, 2005).	NGO
Future of Music Coalition (FMC)	"The Future of Music Coalition is a not-for-profit collaboration between members of the music, technology, public policy and intellectual property law communities. The FMC seeks to educate the media, policymakers, and the public about music / technology issues, while also bringing together diverse voices in an effort to come up with creative solutions to some of the challenges in this space. The FMC also aims to identify and promote innovative business models that will help musicians and citizens to benefit from new technologies. The FMC actualizes its mission through a number of activities. First, we organize public discussion of issues that impact musicians and the public at large, making sure to include a variety of voices in the conversation. Second, we submit testimony, publish articles and speak on panels to make sure the creators' experience is heard. Third, we encourage musicians and citizens to publicly document their experiences on the FMC website. Finally, we generate original research on historic trends and issues of import to the public to more completely illuminate the mechanics of the music industry" (FMC 2005).	NGO

Table A.3
(continued)

Organization	Mission Statement	Class
Our Media	"Ourmedia is a global community and learning center where you can gain visibility for your works of personal media. We'll host your media forever—for free. Video blogs, photo albums, home movies, podcasting, digital art, documentary journalism, home-brew political ads, music videos, audio interviews, digital storytelling, children's tales, Flash animations, student films, mash-ups—all kinds of digital works have begun to flourish as the Internet rises up alongside big media as a place where we'll gather to inform, entertain and astound each other" (Our Media 2005).	NGO-Grass Roots
America Association of Law Libraries	"Working with its Copyright Committee, AALL monitors many legislative, political and judicial developments that affect domestic and international copyright law." (AALL 2005)	NGO
Home Recording Rights Coalition (HRRC)	"HRRC works in Washington, D.C. to protect your right to buy and use audio and video recorders, players, and PCs. Through this site, we provide current information about consumer home recording in the digital age—'hot' topics; past, present, and future congressional activity; FCC [Federal Communications Commission] proceedings; and litigation" (HRRC 2005).	industry/consumer coalition
Association for Computing Machinery	Largest computer professional association in America. "ACM interacts with US government organizations, the computing community, and the public on public policies affecting information technology. Supported by the ACM Office of Public Policy, USACM seeks to inform the U.S. government about policies that impact the computing community and the public. It also identifies significant technical and public policy issues; monitors information on relevant U.S. government activities; and responds to requests for information and technical expertise from U.S. government agencies and departments" (ACM 2005).	NGO

Notes

Chapter 1

1. It is important to note that the term *hacker* is not used here in the fashion popularized by the mass media, which has equated hackers with those who break into computer systems or design malicious computer viruses. The latter are more properly referred to as "crackers" or "phreaks." My use of the term *hackers* is consistent with the meaning ascribed to it by those early authors describing hacker culture, such as Stephen Levy (1984), Eric Raymond (2002), and Richard Stallman (2002): In their definition, hackers are more like tinkerers whose curiosity drive them to "look under the hood" of computer programs and systems and then share what they learn with fellow tinkerers. While this definition simplifies the complexity of hacker culture and the tensions that exist in that conceptualization (often hackers are "looking under the hood" of proprietary systems), for our purposes it suffices because it focuses on the more-often-than-not benign intentions behind true hacking.

2. The DMCA (Pub. L. No. 105-304, 112 Stat. 2860) was enacted in 1998 to bring the United States into compliance with a series of international treaties administered by the World Intellectual Property Organization (WIPO). The DMCA provisions that make copyright-protection circumvention technologies illegal (the anticircumvention provisions) are the center of an ongoing debate that seeks to establish a balance between the rights of consumers and the rights of copyright owners.

3. Some may argue that this is what the movement was always about. Perhaps, but I would suggest that today it is more clearly and uniformly articulated by activists and movement organizations.

4. That logic argues that intellectual-property law was designed primarily to incentivize creators and that without such legal protections creative people would simply stop producing, and our intellectual cache would be impoverished. This logic, espoused primarily by the cultural industries, ignores that, at least in the United States, the framers of the Constitution balanced those protections with considerations for the public domain.

5. Those externalities might include the burden that consumers or future creators must bear when they hope to appropriate existing content or other intellectual properties to engage in their own creative activity.

6. I am always loath to use the term *never before*, but when have consumers actually argued for the right to significantly alter mass-media content outside of the safeguards of "fair use" in other similar doctrines?

7. Americans with Disabilities Act (Pub. L. 101-336, 104 Stat. 327 [1990]).

8. Another analogy might be the act of sit-ins during the civil rights movement in the United States. The sit-ins at "white-only" lunch counters challenged norms and law but also thrust into the white world an alternative view of the world as it should be. The presence of black Americans at white-only establishments created the new world of civil rights before the infrastructures (laws) were there to realize them formally.

9. What I am thinking about when I mention the "normative power of law" is how law through its political power also makes implicit how society ought to be structured. For example, copyright law in the United States and its base-level definitions of authors' rights, which are strong, not only explicitly define the rights but also implicitly suggest that those rights are the ones that *ought* to be given. My idea of hacks in this regard—the alternative-licensing schemes such as Creative Commons and hacks such as DeCSS—put pressure on the normatively implied structure through what they make possible and through the arguments that activists present to legitimize their existence.

10. See, for example, RealNetworks, Inc. v. Streambox, Inc. (US Dist. [W.D. Wash. 2000]) and Universal City Studios, Inc. v. Corley (US Court of Appeals [2d Cir. 2001]), where Streambox and Corley claimed fair use as a defense for circumvention. See Felten et al. v. Recording Industry Association of America, Inc. [RIAA] (US District Court [NJ 2001]), where Felten asked the court to judge on the implication of the DMCA regarding his First Amendment rights and on the impacts of DMCA research and innovation.

11. For example, open anonymous networks versus trusted systems that require personal information and identification.

12. See Strahilevitz 2003 for a recent example of this process in the case of peer-to-peer technologies where the design of some applications is meant to structure use of the application in a way that makes the user more likely to share his or her music.

13. Strong democracy is a highly participatory and decentralized form of democracy espoused by political scientist Benjamin Barber (1984).

14. The concept of "technological resistance technologies" is related to Bryan Pfaffenberger's (1992) concept of the "counterartifact." Pfaffenberger's account of

the narratives that create the politics of technology is parallel to frame analysis in social movement theory. Both approaches suggest that the stories we tell to rationalize the norms, technologies, and laws that govern our behavior are important in coordinating how we act out our dictated roles. Pfaffenberger also uses the concept of "technological regularization," which is analogous to technological enforcement. However, I contend that technology need not have ongoing ritual or narrative to support its political effects. Thus, technological enforcement and regulation are unlike regularization in that they are absolute categories. They do not depend on narrative to exert force. As a consequence, I would argue that technological protection measures (TPMs) are unlike counterartifacts because even when they become designified, they continue to have politics.

15. Andrew Feenberg (1999) draws similar conclusions in *Questioning Technology*. However, his interpretation of technocracy is more traditional than mine: rational, objective analysis of social problems by ruling experts and scientific planning and rationally designed systems would solve ideological problems. Of course, he notes, technologies are ideological themselves, and the objective technocrat is the ghost of a myth.

16. Dale Rose and Stuart Blume (2003), for example, have shown in their study of vaccines that users resist state and scientific configurations through specific acts of dissent or noncompliance.

Chapter 2

1. The Copyright Term Extension Act of 1998 (Pub. L. No. 105-298, 112 Stat. 2827) would further extend copyright to the life of the author plus seventy years. The 1976 act was applicable to works made on or after January 1, 1978. For works in their first term, the copyright would last twenty-eight years from the original start date. For posthumous work or work for hire, the term could be extended to forty-seven years, with the Copyright Term Extension Act extending that term to sixty-seven years.

2. Jessica Litman has written at length concerning the legislative history of the DMCA (see, e.g., Litman 2001), and this section is greatly indebted to her work.

3. See appendix A for a complete list of individuals and organizations that submitted comments.

4. The AHRA (Pub. L. 102-563, 106 Stat. 4237 [1992]) mandated that all players/recorders of digital cassette tapes carry the serial-copy-management system. This technology would limit the number and quality of copies made on digital tape players.

5. Note that these consequences focus on the effects of the NII on copyright; other sections of the Green Paper discussed US obligations with respect to a series of

international treaties, detailed discussion of which is beyond the scope of this work. See Lehman 1994.

6. Copyright Act (Pub. L. 94-553, 90 Stat. 2541 [Oct. 19, 1976]).

7. The Copyright Act defines copies as "material objects, other than phonorecords, in which a work is fixed by any method now known or later developed, and from which the work can be perceived, reproduced, or otherwise communicated, either directly or with the aid of a machine or device" (section 101). The term *copies* encompasses the material object, other than a phonorecord, in which the work is first fixed, and the term *phonorecords* refers to "material objects in which sounds, other than those accompanying a motion picture or other audiovisual work, are fixed by any method now known or later developed, and from which the sounds can be perceived, reproduced, or otherwise communicated, either directly or with the aid of a machine or device. The term 'phonorecords' includes the material object in which the sounds are first fixed" (section 101).

8. Copies are arguably always made, even in the case of streaming content, but these copies are exempted from infringement because they are a consequence of the technology's normal operation.

9. Case law and the legislative history of the Copyright Act had shown that transmission of content among computers resulted in sufficiently permanent fixation. The US House of Representatives report accompanying the Copyright Act noted that digital format is a fixation method covered by the act. In spite of this guidance, the WGIP wanted statutory language to make this point clear.

10. "First sale" is a limitation on the copyright holders' exclusive right to distribute their work codified in section 109 of Title 17, the US copyright law. This section states: "[T]he owner of a particular copy or phonorecord lawfully made under this title, or any person authorized by such owner, is entitled, without the authority of the copyright owner, to sell or otherwise dispose of the possession of that copy or phonorecord."

11. In testimony, critics of the Green Paper noted that this conception explicitly assumes that the consumer is going to keep the original copy, and they chastised copyright owners and the WGIP for assuming that customers will be dishonest.

12. Proponents of expanded use of copyright content argue that courts have left interpretation of the extent of fair use open. See Jessica Litman's and Pamela Samuelson's comments following the release of the Green Paper (*Comments* 1994).

13. The Conference on Fair Use, or CONFU, brought together copyright and user interests to determine the potential impacts of the White Paper recommendations on fair use. Importantly, user interests were mostly represented by institutional actors such as libraries and the measures discussed were primarily aimed at establishing guidelines for libraries and educators.

14. User advocates criticized the recommendation to revisit fair use in the Conference on Fair Use, saying that the issue belonged in rule-making procedures and that the conference was in fact an attempt to marginalize the issue. See the comments by Jessica Litman in *Comments* 1994.

15. The Telecommunications Act of 1996 (Pub. L. No. 104-104, 110 Stat. 56) is a revision of the Communications Act of 1934 (Pub. L. No. 416-652, 48 Stat. 1064).

16. The sale of copying equipment, like the sale of other articles of commerce, does not constitute contributory infringement if the product is widely used for legitimate, unobjectionable purposes. Indeed, it need merely be capable of substantial noninfringing uses.

17. The case of DeCSS, the hack of the DVD access-control technology, is a case in point. Makers of DeCSS (mostly hackers in the open-source movement) failed to convince the courts that the technology was in fact intended for the development of DVD players using the Linux operating system. As such, a host of software developers and users were criminalized. In the five years since this case, it is clear the DeCSS indeed served legitimate noninfringing purposes because the DeCSS technology and its analogs are now found in all software that allows DVDs to be played on Linux machines.

Chapter 3

1. See comments from ISPs such as America Online and AT&T in *Comments* 1994.

2. See, in general, *Comments* 1994.

3. Lawrence Lessig and James Boyle are two other prominent legal scholars who have espoused this view. In general, they believe that digital technologies for editing and sampling can make active cultural remixers of all uses of digital media. See Boyle 1997; Lessig 2004.

4. Recent events such as Grey Tuesday in which more than five hundred Web sites distributed a banned remix of two popular artists' songs as an act of "digital civil disobedience" bear out Mr. Rothman's prediction. See http://www.downhillbattle.org and chapter 8 for more on Grey Tuesday.

5. The NET Act (Pub. L. No. 105-147, 111 Stat. 2678 [1997]) became law a year before the DMCA, so in a sense the White Paper served to buttress the policy rationale for an overall coordinated strategy to ready the DMCA for the realities of the virtual networks.

6. As is evident in the surge in lawsuits brought against consumers following changes in copyright law in the late 1990s—where, for example, a single mother in Chicago was ordered to pay $22,000 for sharing thirty songs on a peer-to-peer network—the extent of criminal liability appears increasingly out of sync with the

crime (Newton 2005a). Consumer surveys in 2005, for example, show that sharing of copyrighted material continued unabated in spite of such lawsuits (Madden and Rainie 2005; Newton 2005b).

7. The Berne Convention for the Protection of Literary and Artistic Works is the oldest and principal treaty regulating copyright internationally.

8. Pamela Samuelson has noted that the DMCA went beyond the obligations.

9. Here I am reading 1201(b) to imply that "manufacture" means manufacture with the intent to distribute. But if read more literally, the statute might mean that even the manufacture by a single individual for her own use is illegal. In which case, the allowance for circumvention of copy-control technology would be even more hollow.

10. A point also raised by Pamela Samuelson (1999).

Chapter 4

1. This feature works as long as the computers are networked and the eBook reader on one computer can check on the originating computer so that multiple instances of the book are not opened. However, most publishers will not allow for this sharing. The Microsoft Reader does not have this feature and allows you to read the book in only two authorized machines.

2. I discuss Creative Commons and the Creative Commons license in chapter 8.

3. The DeCSS case (reviewed in chapters 5 and 6) also made the same argument. The argument that code is speech had been made in previous cases. See Bernstein v. US Department of State, 922 F. Supp. 1426 (ND Calif. 1996).

4. My summary of ElcomSoft's arguments is derived from transcripts of the cases United States v. Dmitry Sklyarov and United States v. ElcomSoft, US District Court (ND Calif., San Jose Div., 2001 and 2002).

5. ElcomSoft in fact made this comparison explicit as it cited Chicago Lock Co. v. Fanberg (676 F.2d 400, 216 USPQ [BNA] 289 [9th Cir. 1982), a case involving instructions on how to reverse engineer locks protected by trade secret.

6. This suggestion was made in pre–September 11, 2001 days, when tracking of every citizen was generally an abhorrent idea to the US public. Sentiments may have changed since then.

7. ElcomSoft noted: "The DMCA does not purport to prohibit the violation of copyright laws. Instead, it regulates speech that might facilitate such violations. . . . Congress focused not on the infringer himself, but rather on a person more removed from the infringement (but perhaps easier to locate). However expedient such an approach might be, it fails to maintain the crucial connection between the

government's ends and the means used to accomplish them. The infringement of a copyright is wrong in and of itself; the circumvention of a technological measure protecting that copyright is only wrong in those circumstances in which a copyright will be infringed as a result" (United States v. ElcomSoft [2002], Motion to Dismiss on First Amendment Grounds).

8. These organizations included the EFF, Association for Computing Machinery, American Association of Law Libraries, Consumer Project on Technology, Electronic Privacy Information Center, and Music Library Association (United States v. Dmitry Sklyarov [2001], Amicus Brief EFF et al.).

9. "When speech and non-speech elements are combined in a single course of conduct, a sufficiently important government interest in regulating the non-speech element can justify incidental intrusions on First Amendment freedoms" (United States v. ElcomSoft [2002], Motion to Dismiss on First Amendment Grounds).

10. "The principal inquiry in determining whether a statute is content-neutral is whether the government has adopted a regulation of speech because of agreement or disagreement with the message it conveys" (United States v. ElcomSoft [2002]).

Chapter 5

1. Macrovision is the copy-protection technology mandated by Congress on all VHS tape recorders in response to copyright owners' fears that films on VHS tapes would be copied "en masse" by consumers.

2. Notably, reverse engineering of software applications for the purposes of interoperability and education is permissible under exemptions of the copyright law. The license asks users to give up this privilege as a condition of use.

3. "There is no legal precedent or court decision in Norway to support a claim that reverse engineering is a violation of Norwegian criminal law. No Norwegian court has issued any such ruling" (DVD CCA v. Bunner et al. [1999], Declaration of Jon Bing for Defendants).

4. "The ability of the software at issue to play DVD discs from various regions does not violate any right or privilege available under law to the copyright owner of the movie on the disc; 'code-free' consumer DVD players already exist and offer the same capability. In my opinion, the regional coding system was built as a business strategy, to give a technological edge to theater owners, to the disadvantage of consumers; there are no legal consequences if this intended edge does not materialize in practice" (DVD-CCA v. Bunner et al. [1999], Declaration of John Gilmore).

5. These data are derived from analysis of subsequent filings, in which some defendants were dropped because they complied with the court order, whereas the great majority did not.

Chapter 6

1. Review of the transcripts from this hearing shows that Judge Lewis Kaplan allowed the defense very little latitude, questioning every motion and argument, sometimes without giving the EFF a chance to complete its argument. At times, the defense is demonstrably flustered by Kaplan's antagonism. In contrast, the plaintiffs received no such treatment or questioning; in fact, Judge Kaplan simply accepted their arguments and became their most vocal advocate, even submitting a well-prepared explanation of his decision to grant the injunction during the hearing.

2. In context, many Internet posters were expressing frustration at being banned from distributing software that they felt represented free speech and the ability to make Linux DVD players. However, in court the rhetoric became indefensible.

3. Garbus has defended Lenny Bruce, Henry Miller, and Salman Rushdie, among others, in First Amendment cases.

4. Readers' ability to access primary materials has given rise to citizen journalism through blogs and thus in turn ironically challenged the institutional authority of mainstream media news.

5. Institutional tactics are methods of challenging law and repression through traditional structures such as courts or the legislature. Extrainstitutional tactics are outside of established structures and include disruptive collective action such as protests, civil disobedience, and hacking.

6. According to section 1201(f) of the DMCA, "§Notwithstanding the provisions of subsection (a)(1)(A), a person who has lawfully obtained the right to use a copy of a computer program may circumvent a technological measure that effectively controls access to a particular portion of that program for the sole purpose of identifying and analyzing those elements of the program that are necessary to achieve interoperability of an independently created computer program with other programs, and that have not previously been readily available to the person engaging in the circumvention, to the extent that any such acts of identification and analysis do not constitute infringement under this title."

7. For example, a report from the Pew Internet and American Life Project noted that after the RIAA started suing individual file swappers on peer-to-peer networks, the number of users dropped by almost half. See Rainie and Madden 2004.

8. For example, the US Supreme Court decision in MGM v. Grokster noted that the technology makers would be held accountable if there was no legitimate use *and* if there was inducement on the part of the technology makers to violate copyright (MGM Studios et. al v. Grokster, LTD., et. al, Case No. 04-480 June 27,

2005). Both sides claimed this decision a victory. The makers of peer-to-peer technology, for example, noted that as long as they make it clear that their technology is meant for sharing of legal files, then they can go about their business.

Chapter 7

1. By May 7, a little more than a week after launch, Apple sold one million songs on the iTMS. See Wilcox 2003.

2. Apple continuously tweaks its EULA, so since the time of this writing some changes may have occurred. However, the EULA's basic mechanics and general thrust remain the same. As noted later in this chapter, the TOSA and EULA have changed so that many of the restrictions talked about here have been lifted, particularly on music downloads.

3. As the business model took hold, Apple lowered its restrictions to five songs. Today these restrictions have been lifted, and music bought on the iTMS is no longer protected by DRM.

4. Of course, content on a computer is always at some level accessible as a copy either through the analog hole or by capturing the data on its way to the software that plays it. Secure hardware architectures promise to plug these gaps.

5. The services were formerly available at Spymac.com and ShareiTunes.com. These applications are no longer available.

6. See Ian Freid quoting an Apple release at http://news.com.com/Apple+limits+iTunes+file+sharing/2100-1027_3-1010541.html.

7. This type of work-around was later stymied by the release of iTunes 4.5, which prevented compatibility between various version of iTunes.

8. These sorts of statements concerning the legality of a technology tell us much about user attitudes concerning what users ought to be able to do with legally purchased content. For users, issues of access and personalization (which, for the digital rights movement, are translated into issues of creative and participatory privileges) continue to be central.

9. MyTunes resurfaced in September 2004 under the name "MyTunes Redux." See http://minimalverbosity.com and Borland 2004a.

10. Although Johansen's involvement was integral in the development of DeCSS, it was not fleshed out in chapter 5 or 6 because the analysis there focused on DeCSS in the US courts. Johansen is a Norwegian citizen, and although he was tried in Norway for his development of DeCSS, he did not play a large role in the US court cases.

11. Many legal scholars have commented that DMCA section 1201 essentially grants a new right to copyright owners, the "access right."

12. During development of an open-source project such as VLC, contributors will add their contributions to a public library of applications, which serves as an unofficial collection of proposed changes to the project. Project leaders will periodically update the official release of the application with work from the unofficial library.

13. BIOS stands for "basic input/output system." It is a software package written for a computer's bootable memory (flash chip, ROM, or RAM). The BIOS allows a PC user to have access to input/output devices such as monitors and keyboards before the operating system has loaded.

14. Note that when a song is transferred to an unauthorized computer, it has the encryption from the authorized computer. Thus, VLC on the unauthorized computer cannot simply crack it as it did on the authorized computer. That is why the keys generated from VLC's cracking on the authorized machine must also be transferred to the unauthorized machine.

15. Recall that the VLC component did not produce an unprotected file; it generated the song keys, which could be copied along with the DRM-protected song to unauthorized machines that used the VLC. The VLC installed on unauthorized machines would use the imported song keys to play the DRM-protected song.

16. The Hymn Project Web page contradicts the "readme" file, stating that there are no GUIs available for Mac, yet the "readme" on the Hymn 7.0 release shows the GUI. Thus, I assume that the early version did not have a GUI and that the later version—probably when PlayFair became Hymn—does have a Mac GUI. Also note that the song keys must still be generated with an iPod, so for PlayFair to work on a Mac, the iPod with those songs must be connected.

17. I am assuming that the developer is male based on a review of interviews with media, postings on the Hymn Web site, and the developer's references to himself.

18. Johansen also designed PodKey, a utility that would access an iPod's "key store." PodKey can retrieve the song keys from an iPod and make them available to DeDRMS. Thus, a user without an Internet connection to the iTMS server can still authorize more than one computer to play a song by simply having her iPod handy. This is similar to what PlayFair/Hymn does when used on a Mac.

19. A jury might find that these technologies' self-policing features show that the developer did not intend to facilitate egregious file sharing, but only to allow limited file copying for loosely defined personal uses.

Chapter 8

1. Pub. L. No. 103-414, 108 Stat. 4279.

2. The reader will notice that in my discussion of Creative Commons I do not address its framing tactics in the media. This omission has been made because Cre-

ative Commons has not done media outreach to a large extent. Although it does get its message "out" via traditional media outlets, its primary audience is artists and other intellectual-property creators whose adoption of Creative Commons licenses is sought by the organization. Therefore, framing strategies are tailored toward these niche groups and typically do not find their way into mass media.

3. At the time that Battle Labs was surveyed, the projects were under various stages in development. They are listed here as an example of the kinds of technological measures that were potential parts of the digital rights movement.

References

American Libraries Association. At http://www.ala.org/advocacy/copyright. Retrieved June 2005.

American Association of Law Libraries. At http://www.aallnet.org/main-menu/Advocacy/copyright. Retrieved February 2005.

Aogail. 2004. Re: "Legitimate" Uses. At http://macslash.org/comments.pl?sid=4434&op=&threshold=0&commentsort=0&mode=thread&pid=70578#70581. Retrieved February 22, 2005.

Association for Computing Machinery. At http://www.acm.org/public-policy. Retrieved February 2012.

Banks, J. 2005. Opening the Pipeline: Unruly Creators. Paper presented at the Digital Games Research Association, Vancouver, CA, June.

Barber, B. R. 1984. *Strong Democracy: Participatory Politics for a New Age*. Berkeley and Los Angeles: University of California Press.

Barlow, J. P. 1996. A Declaration of the Independence of Cyberspace. At http://homes.eff.org/~barlow/Declaration-Final.html. Retrieved January 2006.

Barlow, J. P. 1994. The Economy of Ideas. *Wired* 2 (3). At http://www.wired.com/wired/archive/2.03/economy.ideas.html. Retrieved February 2012.

Biegel, S. 2001. *Beyond Our Control? Confronting the Limits of Our Legal System in the Age of Cyberspace*. Cambridge, MA: MIT Press.

Borland, J. 2003a. Apple Unveils Music Store. At http://news.com.com/Apple+unveils+music+store/2100-1027_3-998590.html. Retrieved February 9, 2005.

Borland, J. 2003b. Hackers: iTunes Can Be Shared over Net. At http://news.com.com/Hackers+iTunes+can+be+shared+over+Net/2100-1027_3-1001121.html. Retrieved February 9, 2005.

Borland, J. 2003c. iTunes Helper Allows MP3 Downloads. At http://news.com.com/iTunes+helper+allows+MP3+downloads/2100-1027_3-5107196.html. Retrieved February 10, 2005.

Borland, J. 2004a. Hackers Revive iTunes Music Sharing. At http://news.com.com/Hackers+revive+iTunes+music+sharing/2100-1026_3-5316700.html. Retrieved February 10, 2005.

Borland, J. 2004b. iTunes Song Swap Helper Vanishes from Net. At http://news.com.com/iTunes+song+swap+helper+vanishes+from+Net/2100-1027_3-5171519.html. Retrieved February 10, 2005.

Boyle, J. 1997. A Politics of Intellectual Property: Environmentalism for the Net? At http://www.law.duke.edu/boylesite/intprop.htm. Retrieved June 2005.

Boyle, J. 1996. *Shamans, Software, and Spleens: Law and the Construction of the Information Society*. Cambridge, MA: Harvard University Press.

Brown, R. H. 1992. The National Information Infrastructure: Agenda for Action. At http://www.ibiblio.org/nii/toc.html. Retrieved January 2006.

CDT (Center for Democracy and Technology). 2005. About Us. At http://www.cdt.org. Retrieved June 2005.

Chaosmint. 2003. QTFairUse, MyTunes, and Protected AAC Explained: Don't Believe Everything You Read. At http://www.chaosmint.com/macintosh/articles/qtfairuse-mytunes-itunes.shtml. Retrieved February 15, 2005.

Chilling Effects: n.d. Cease-and-Desist Clearing House at http://www.chillingeffects.org. Retrieved June 2005.

CIS (Center for Internet and Society), Stanford University. 2005. About Us. At http://cyberlaw.stanford.edu. Retrieved June 2005.

Cohen, J. 2005. The Place of the User in Copyright Law. *Fordham Law Review* 74 (1): 347–374.

Cohen, P. 2003. Apple Releases iTunes 4, QT 6.2. iPod Update 1.3. At http://www.macworld.com/news/2003/04/28/software. Retrieved February 7, 2005.

Cohn, C. 2003. California Supreme Court Upholds Free Speech in DVD Case. At http://www.eff.org/IP/Video/DVDCCA_case/20030825_eff_bunner_pr.php. Retrieved June 2005.

Comments Submitted for the Public Hearing on Intellectual Property Issues Involved in the National Information Infrastructure Initiative, Andrew Mellon Auditorium, Washington, DC. 1993. US Patent and Trademark Office, US Department of Commerce. At http://www-personal.umich.edu/~jdlitman/NOV18NII.TXT. Retrieved February 2012.

Comments Submitted for the Public Hearing of the National Information Infrastructure Task Force, Washington, DC; Los Angeles, Chicago. 1994. US Patent and Trademark Office, US Department of Commerce. At http://w2.eff.org/IP/ipwg_nii_ip_report_acis.comments.txt. Retrieved February 2012.

Copyfight. 2005. About Us. At http://copyfight.corante.com. Retrieved June 2005.

CPSR (Computer Professionals for Social Responsibility). n.d. At http://cpsr.org/about. Retrieved June 2005.

Creative Commons. 2005. About Us. At http://www.creativecommons.org. Retrieved September 2005.

Dalrymple, J. 2003. MyTunes Exploits iTunes Windows Playlist Sharing. At http://www.macworld.com/news/2003/11/12/mytunes. Retrieved February 9, 2005.

DFC (Digital Future Coalition). n.d. At http://www.artistsresourceguide.org/Digital_future_coalition. Retrieved June 2005.

Downhill Battle. 2005a. About Us. At http://www.downhillbattle.org. Retrieved August 2005.

Downhill Battle. 2005b. Downhill Battle Labs: Strategic Software for Music Activism. At http://www.downhillbattle.org/labs. Retrieved August 2005.

Downhill Battle. 2004. Grey Tuesday. At http://www.greytuesday.org. Retrieved June 2005.

DVD Encryption Cracked. 1999. *2600: The Hacker Quarterly*. At http://www.2600.com/news/view/article/20. Retrieved June 2005.

Earl, J., and A. Schussman. 2003. The New Site of Activism: On-line Organizations, Movement Entrepreneurs, and the Changing Location of Social Movement Decision Making. *Research in Social Movements, Conflicts, and Change* 22:155–187.

EFF (Electronic Frontier Foundation). 2005. About Us. At http://www.eff.org. Retrieved June 2005.

EFF (Electronic Frontier Foundation). 2001a. EFF "Intellectual Property: Digital Millennium Copyright Act (DMCA): U.S. v. ElcomSoft & Sklyarov" Archive. At http://www.eff.org/IP/DMCA/US_v_Elcomsoft/. Retrieved June 2005.

EFF (Electronic Frontier Foundation). 2001b. FBI Arrests Programmer in Las Vegas: Distributed Tool That Increases Purchaser's Control of eBooks. At http://www.eff.org/IP/DMCA/US_v_Elcomsoft/20010717_eff_sklyarov_pr.html. Retrieved June 2005.

EFF (Electronic Frontier Foundation). 2001c. Key Legislators on Fair Use and DMCA: Boucher & Ashcroft Speak against Criminalization of Legitimate Software. At http://www.eff.org/IP/DMCA/US_v_Elcomsoft/boucher_ashcroft_dmca.html. Retrieved June 2005.

EFF (Electronic Frontier Foundation). 2001d. Protest Adobe's Role in Jailing Programmer Sklyarov. At http://www.eff.org/IP/DMCA/US_v_Elcomsoft/20010719_eff_sklyarov_pr.html. Retrieved June 2005.

EPIC (Electronic Privacy Information Center). 2005. About Us. At http://www.epic.org. Retrieved June 2005.

Fallout from Def Con: Ebook Hacker Arrested by FBI. 2001. At http://yro.slashdot.org/comments.pl?sid=13272&threshold=1&mode=nested&commentsort=0&op=Change. Retrieved June 2005.

Feenberg, A. (1999). *Questioning Technology*. New York: Routledge.

FMC (Future of Music Coalition). n.d. At http://futureofmusic.org/about. Retrieved June 2005.

Fried, I. 2003a. Apple Limits iTunes File Sharing. At http://news.com.com/Apple+limits+iTunes+file+sharing/2100-1027_3-1010541.html. Retrieved February 11, 2005.

Fried, I. 2003b. Music-Swapping Software Makes Comeback. At http://news.com.com/Music-swapping+software+makes+comeback/2100-1027_3-997039.html. Retrieved February 10, 2005.

FSF (Free Software Foundation). 2005. About Us. At http://www.fsf.org/about. Retrieved June 2005.

FutureProof. 2004. JHymn Info and Help. At http://www.hymn-project.org/jhymndoc/#whatsnew. Retrieved February 26, 2005.

Geekpatrol. 2003. QTFairUse in Use. At http://www.geekpatrol.ca/archives/2003/12/08/qtfairuseinuse.php. Retrieved February 16, 2005.

GILC (Global Internet Liberty Campaign). 2005. About Us. At http://www.gilc.org. Retrieved June 2005.

Gillespie, T. 2007. *Wired Shut: Copyright and the Shape of Digital Culture*. Cambridge, MA: MIT Press.

Greenman, B. 2004. The Mouse That Remixed. *The New Yorker*, February 9.

Hartley, J. 2006. Facilitating the Creative Citizen. At http://www.onlineopinion.com.au/view.asp?article=5036. Retrieved November 2006.

Heidi, J. 2003. iTunesDL and iSlurp: Downloading via iTunes Sharing. At http://www.macilife.com/2003_05_11_archive.html. Retrieved February 11, 2003.

Hess, D., & Martin, B. (2006). Backfire, Repression, and the Theory of Transformative Events. *Mobilization. International Quarterly*, 11(2), 249–267.

Holcomb, G. 2003. iSlurp b3. At http://www.oatbit.com/iSlurp. Retrieved February 12, 2005.

HRRC (Homer Recording Rights Coalition). n.d. At http://www.hrrc.org. Retrieved June 2005.

Hymn Project. 2004a. Home Page. At http://hymn-project.org. Retrieved February 2005.

Hymn Project. 2004b. Hymn-ReadMe (Version 0.7.1). At Hymn-project.org.

Hymn Project. 2004c. Hymn User Manual. At http://www.hymn-project.org/docs/hymn-manual.html. Retrieved February 27, 2005.

Hymn Project. 2004d. iOpener FAQ. On iOpener (Version 0.2).

Hymn Project. 2004e. Old News. At http://hymn-project.org/forums/viewtopic.php?t=3. Retrieved February 22, 2005.

Illegal Art Audio. 2004. At http://www.illegal-art.org/audio/liner.html. Retrieved November 2005.

iTunes 4 Tip—Sharing iTunes Libraries over IP; It's Not Just for Rendezvous. 2003. *MacObserver*. At http://macobserver.com/article/2003/04/28.14.shtml. Retrieved February 11, 2003.

iTunesDL. 2003a. *MacUpdate*. At http://www.macupdate.com/app/mac/5661/apple-itunes. Retrieved February 12, 2005.

iTunes Tracker 1.0. 2003b. *MacUpdate*. At http://www.macupdate.com/info.php/id/12088. Retrieved February 11, 2005.

Jailed under a Bad Law (editorial). 2001. *Washington Post*, August 21.

Jenkins, H. 1992. *Textual Poachers: Television Fans & Participatory Culture*. New York: Routledge.

Johansen, J. 2004. "Readme" for DeDRMS (Version 0.1).

Jordan, T. 2002. *Activism! Direct Action, Hacktivism, and the Future of Society*. London: Reaktion Books.

Jordan, T., and P. A. Taylor. 2004. *Hacktivism and Cyberwars: Rebels with a Cause?* 1st ed. New York: Routledge.

Katalov, V. 2001. Advanced Acrobat eBooks Are NOT Really Secure. At http://www.planetpdf.com/mainpage.asp?webpageid=2393. Retrieved June 2005.

Khaney, L. 2003. iTunes Music Swap Just Won't Die. At http://www.wired.com/news/mac/0,2125,59127,00.html?tw=wn_story_related. Retrieved February 14, 2005.

Kucklich, J. 2005. Precarious Playbour: Modders and the Digital Games Industry. *Fibreculture* 5. At http://journal.fibreculture.org/issue5/kucklich_print.html. Retrieved September 2005.

Lasica, J. D. 2005. *Darknet: Hollywood's War against the Digital Generation*. Hoboken, NJ: Wiley.

Lehman, B. 1994. *Intellectual Property and the National Information Infrastructure: A Preliminary Draft of the Report of the Working Group on Intellectual Property Rights*. Washington DC. At http://cool.conservation-us.org/bytopic/intprop/ipwg. Retrieved February 2012.

Lessig, L. 2004. *Free Culture: How Big Media Uses Technology and the Law to Lock Down Culture and Control Creativity*. New York: Penguin.

Lessig, L. 2001. Jail Time in the Digital Age. *New York Times*, July 30.

Lessig, L. 1999. *Code and Other Laws of Cyberspace*. New York: Basic Books.

Levy, S. 1984. *Hackers: Heroes of the Computer Revolution*. New York: Penguin.

Lindsay, C. 2003. From the Shadows: Users as Designers, Producers, Marketers, Distributors, and Technical Support. In N. Oudshoorn and T. J. Pinch, eds., *How Users Matter: The Co-Construction of Users and Technologies*, 29–50. Cambridge, MA: MIT Press.

Litman, J. 2001. *Digital Copyright: Protecting Intellectual Property on the Internet*. Amherst, NY: Prometheus Books.

Localman. 2004. Re: Overseas. At http://macslash.org/comments.pl?sid=4434&op=&threshold=0&commentsort=0&mode=thread&pid=70575#70578. Retrieved February 22, 2005.

Madden, M., and L. Rainie. 2005. *Music and Video Downloading Moves Beyond*. Washington, DC: Pew Internet and American Life Project.

McAdam, D., J. D. McCarthy, and M. N. Zald. 1996. *Comparative Perspectives on Social Movements: Political Opportunities, Mobilizing Structures, and Cultural Framings*. New York: Cambridge University Press.

Menta, R. 2003. MyTunes Turns iTunes into File Trading Service. At http://www.mp3newswire.net/stories/2003/mytunes.html. Retrieved February 12, 2005.

Newton, J. 2005a. Pay $22.5K, RIAA p2p Victim Told. At http://www.p2pnet.net/story/7273. Retrieved November 2005.

Newton, J. 2005b. Software Piracy Is Rampant. At http://p2pnet.net/story/7267. Retrieved January 3, 2006.

Orlowski, A. 2004. iTunes DRM Cracked Wide Open for GNU/Linux. Seriously. *The Register*, January 5. At http://www.theregister.co.uk/2004/01/05/itunes_drm_cracked_wide_open. Retrieved February 18, 2004.

Orlowski, A. 2003. There's a Noose in the Hoose—iTunes Shoppers Discover DRM. At http://www.theregister.co.uk/2003/12/02/theres_a_noose/. Retrieved February 15, 2005.

Oudshoorn, N., and T. J. Pinch. 2003. *How Users Matter: The Co-Construction of Users and Technologies*. Cambridge, MA: MIT Press.

Our Media. At http://www.ourmedia.org/node/4972. Retrieved August 2005.

Participatory Culture. n.d. At http://www.pculture.org/pcf/about. Retrieved June 2005.

Patterson, L. R. 1968. *Copyright in Historical Perspective*. Nashville, TN: Vanderbilt University Press.

Pfaffenberger, B. 1992. Technological Dramas. *Science, Technology, & Human Values* 17 (3): 282–312.

Planet eBook. 2001a. Adobe Acrobat eBook Reader Updated after Security Issue: Adobe Systems Threatens Legal Action for Alleged Copyright Infringement. At http://www.planetebook.com/mainpage.asp?webpageid=157. Retrieved June 2005.

Planet eBook. 2001b. Adobe and Electronic Frontier Foundation Want Dmitry Sklyarov Released. At http://www.planetebook.com/mainpage.asp?webpageid=191. Retrieved June 2005.

Planet PDF. 2001. Index of ElcomSoft, Dmitry Sklyarov, Adobe, US Government, and DMCA-Related Coverage. At http://www.planetpdf.com/mainpage.asp?webpageid=2365. Retrieved June 2005.

Postigo, H. 2010. Information Communication Technologies and Framing for Backfire in the Digital Rights Movement: The Case of Dmitry Sklyarov's Advanced e-Book Processor. *Social Science Computer Review* 28 (2): 232–250.

Postigo, H. 2008. Video Game Appropriation through Modifications: Attitudes Concerning Intellectual Property among Modders and Fans. *Convergence: International Journal of Research into New Media Technologies* 14 (1): 59–74.

Postigo, H. 2007. Of Mods and Modders: Chasing Down the Value of Fan Based Digital Game Modifications. *Games and Culture* 2 (4): 300–312.

PRC (Privacy Rights Clearinghouse). At http://www.privacyrights.org/about_us.htm. Retrieved June 2005.

Public Knowledge. 2005. About Us. At http://www.publicknowledge.org. Retrieved June 2005.

QTFairUse? 2003. *MacRumors*. At http://forums.macrumors.com/showthread.php?t=48482. Retrieved February 15, 2005.

Rainie, L., & Madden, M. (2004). *The Impact of Recording Industry Suits against Music File Swappers*. Washington, DC: Pew Internet and American Life Project.

Raymond, E., ed. 2001. *The Cathedral & the Bazaar: Musings on Linux and Open Source by an Accidental Revolutionary*. Sebastopol, CA: O'Reilly Media.

Re: Overseas. 2004. At http://macslash.org/comments.pl?sid=4434&op=&threshold=0&commentsort=0&mode=thread&pid=70575#70578. Retrieved February 22, 2005.

Rose, D., & Blume, S. 2003. Citizens as Users of Technology: An Exploratory Study of Vaccines and Vaccination. In N. Oudshoorn and T. J. Pinch, eds., *How Users Matter: The Co-Construction of Users and Technologies*, 103–132. Cambridge, MA: MIT Press.

Samuelson, P. 1999. Intellectual Property and the Digital Economy: Why the Anti-Circumvention Regulations Need to Be Revised. *Berkeley Technology Law Journal* 14 (519): 46. At people.ischool.berkeley.edu/~pam/papers/dmcapaper.pdf. Retrieved February 2012.

Samuelson Law, Technology, and Public Policy Clinic, Berkeley Boalt School of Law. 2005. About Us. At http://www.law.berkeley.edu/clinics/samuelson. Retrieved June 2005.

Sarovar. 2004. PlayFair Has Been Taken Down. At http://sarovar.org/forum/forum.php?forum_id=474. Retrieved February 20, 2005.

Sclove, R. 1995. *Democracy and Technology*. New York: Guilford Press.

Shirky, C. 2008. *Here Comes Everybody: The Power of Organizing without Organizations*. New York: Penguin.

Sperberg, R. 2001. The Adobe Security Imbroglio: Roger Sperberg Looks at Russian Software That Decrypts Purchased Adobe eBook Files. Part 2. At http://www.ebookweb.org/opinion/roger.sperberg.20010715.aebpr.htm. Retrieved June 2005.

Speth, J. G. 2005. iCommune. At http://icommune.sourceforge.net. Retrieved February 9, 2005.

Stallman, R. 2002. *Free Software Free Society: Selected Essays of Richard Stallman*. Boston: GNU Press.

Steele, S. 2001. EFF Letter from Executive Director Shari Steele to Attorney General John Ashcroft. July 20. At http://www.eff.org/IP/DMCA/US_v_Elcomsoft/20010720_eff_ashcroft_letter.html. Retrieved June 2005.

Strahilevitz, L. J. 2003. Charismatic Code, Social Norms, and the Emergence of Cooperation on the File-Swapping Networks. *Virginia Law Review* 89 (3): 505–595.

Students for Free Culture. n.d. At http://freeculture.org/about. Retrieved June 2005.

Tarrow, S. G. 1994. *Power in Movement: Social Movements, Collective Action, and Politics*. New York: Cambridge University Press.

Testimony, Public Hearing of the National Information Infrastructure Task Force Working Group on Intellectual Property, Andrew Mellon Auditorium, Washington, DC. 1994. US Patent and Trademark Office, US Department of Commerce.

Touretzky, D. S. n.d. Gallery of CSS Descramblers. At http://www.cs.cmu.edu/~dst/DeCSS/Gallery/index.html. Retrieved July 2005.

Triplehorn, G. 2004a. GetTunes 2.3.6. At http://www.macupdate.com/info.php/id/13738. Retrieved February 14, 2005.

Triplehorn, G. 2004b. Project: getTunes: Summary. At http://sourceforge.net/projects/gettunes. Retrieved February 14, 2005.

US Copyright Office. 2001. *DMCA Section 104 Report*. Washington, DC: US Copyright Office.

US Copyright Office. (1998). *The Digital Millennium Copyright Act of 1998: U.S. Copyright Office Summary*. Washington, DC: US Copyright Office.

Warren, R. 2005. The Openlaw DVD/DeCSS Forum: Frequently Asked Questions. At http://eon.law.harvard.edu/openlaw/DVD/dvd-discuss-faq.html. Retrieved May 2005.

Werde, B. 2004. Defiant Downloads Rise from Underground. *New York Times*, February 25, 2004.

WGIP (Working Group on Intellectual Property Rights). 1995. *Intellectual Property and the National Information Infrastructure: The Report of the Working Group on Intellectual Property Rights*. Washington, DC: Information Infrastructure Task Force.

WGIP (Working Group on Intellectual Property Rights). 1994. *Intellectual Property and the National Information Infrastructure: A Preliminary Draft of the Report of the Working Group on Intellectual Property Rights*. Washington, DC: Information Infrastructure Task Force. At http://www.google.com/url?sa=t&rct=j&q=nii+report+copyright&source=web&cd=2&ved=0CC0QFjAB&url=http%3A%2F%2Fwww.uspto.gov%2Fweb%2Foffices%2Fcom%2Fdoc%2Fipnii%2Fipnii.pdf&ei=TWxST5ekAsbwgge66MCgAQ&usg=AFQjCNHkNguvSlluSxgN-R0ELYDGT8mR3Q. Retrieved February 2012.

White, D. 2003. iLeech Home. At http://ileech.sourceforge.net. Retrieved February 12, 2005.

Wilcox, J. 2003. iTunes store: More Than 1 million Sold. At http://news.com.com/iTunes+store+More+than+1+million+sold/2100-1027_3-999701.html. Retrieved February 10, 2005.

Winner, L. 1985. Do Artifacts Have Politics? In D. McKenzie and J. Wajcman, eds., *The Social Shaping of Technology*, 28–40. London: Open University Press.

Index

Page numbers followed by f or t refer to figures and tables, respectively.